Encyclopaedia of

Sustainable Water Resource Management

Encyclopaedia of
Sustainable Water Resource Management

Volume 2

Editor:
R.N. Sahay

AGROTECH PRESS

JAIPUR ▪ NEW DELHI

AGROTECH PRESS

Regd. Off. "Bony Residency", Gate No. 2
Opp. Tilak Public School, Vishweswariya Nagar,
Gopalpura Road, Jaipur - 302018, India
Ph: 0141 - 2761280, Telefax: 2761381
E-mail : agrotechpress@gmail.com

Delhi Office: 102, 1st Floor, Satyam House,
4327/3, Ansari Road, Darya Ganj,
New Delhi - 110002, India
Ph: 011-45652440
visit us: www.agrotechpress.com

Distributed By
KEYA INTERNATIONAL PRESS LLP
E-mail: keyainternationalpressllp@gmail.com

Published in India, 2025

ISBN : 9789384568269 (5 Vols. Set)

Typeset by Thomson Press

Printed and bound in India by Thomson Press Ltd.,
New Delhi

Contents

Preface

Groundwater is water located beneath the earth's surface in soil pore spaces and in the fractures of rock formations. A unit of rock or an unconsolidated deposit is called an aquifer when it can yield a usable quantity of water. The depth at which soil pore spaces or fractures and voids in rock become completely saturated with water is called the water table. Groundwater is recharged from, and eventually flows to, the surface naturally; natural discharge often occurs at springs and seeps, and can form oases or wetlands. Groundwater is also often withdrawn for agricultural, municipal and industrial use by constructing and operating extraction wells. The study of the distribution and movement of groundwater is hydrogeology, also called groundwater hydrology.

Typically, groundwater is thought of as liquid water flowing through shallow aquifers, but technically it can also include soil moisture, permafrost, immobile water in very low permeability bedrock, and deep geothermal or oil formation water. Groundwater is hypothesized to provide lubrication that can possibly influence the movement of faults. It is likely that much of the Earth's subsurface contains some water, which may be mixed with other fluids in some instances. Groundwater may not be confined only to the Earth. The formation of some of the landforms observed on Mars may have been influenced by groundwater. There is also evidence that liquid water may also exist in the subsurface of Jupiter's moon Europa. Certain problems have beset the use of groundwater around the world. Just as river waters have been over-used and polluted in many parts of the world, so too have aquifers. The big difference is that aquifers are out of sight. The other major problem is that water management agencies, when calculating the "sustainable yield" of aquifer and river water, have often counted the same water twice, once in the aquifer, and once in its connected river. This problem, although understood for centuries, has persisted, partly through inertia within government agencies. In Australia, for example, prior to the statutory reforms initiated by the Council of Australian Governments water reform framework in the 1990s, many Australian States managed groundwater and surface water through separate government agencies, an approach beset by rivalry and poor communication. Groundwater is a highly useful and often abundant resource. However, over-use, or overdraft, can cause major problems to human users and to the environment. The most evident problem is a lowering of the water table beyond the reach of existing wells. Wells must consequently be deepened to reach the groundwater; in some places the water table has dropped hundreds of feet because of extensive well pumping. In the Punjab region of India, for example, groundwater levels have dropped 10 meters since 1979, and the rate of depletion is accelerating. A lowered water table may, in turn, cause other problems such as groundwater-related subsidence and saltwater intrusion. Irrigated agriculture is a big business in Asia. It accounts for 70 per cent of the world's irrigated land and 73 per cent of water used each year by the farming industry. As ageing large-scale surface irrigation schemes have become increasingly inefficient, and farmers have begun growing a wider range of crops requiring water on demand, the number of groundwater wells in India has exploded. In 1960, there were fewer than 100,000 such wells; by 2006 the figure had risen to nearly 12 million In India, a possible solution to over-use of groundwater is emerging, known as 'groundwater recharge'. It involves capturing rainwater that would otherwise run-off, and using it to refill aquifers. Since 2000, the International Water Management Institute has been working with the Indian authorities to help improve the availability of water for agriculture in India. In 2006, India's finance minister invited IWMI to submit policy recommendations based on its research on groundwater depletion. One of the key recommendations was to instigate a programme of recharging groundwater across the 65 per cent of India that has hard-rock aquifers. As a result, the Indian government allocated Rs. 1800 crore to fund dug-well recharge projects 100 districts within seven states where water stored in hard-rock aquifers has been over-exploited.

The book provides a platform to bring together earth scientists, professionals from chemical and engineering science disciplines, public health professionals and social scientists involved in the management and development of groundwater resources. The book is expected to reflect the current understanding of all the issues related to management of groundwater resources and their sustainable use. The book shall prove to be useful to hydrogeologists, civil and irrigation engineers, town planners and professionals involved in the related fields.

—Editor

1

Water Resources and Distribution Systems

It is appropriate to examine the concept of water resource systems as a foundation for consideration of the distribution issues that impact human use of the resource throughout the world. One helpful approach is to consider a basic problem in the provision of water, namely the disparity between the distribution of available water versus the demand for the resource in a specific region or locale (Buras, 1972). There are three basic types of maldistribution of water resources. First, one may envision a geographical/spatial maldistribution of the resource; the geographical area may be experiencing a flood condition (excess water) or a drought condition (insufficient water).

In either case, the water availability fails to match the present demand. A second type of maldistribution of water is temporal. For example, the water may be available during the spring of the year following snowmelt in upstream areas. However, this water may not be available later in the growing season when it is needed for irrigation purp oses. The third type of maldistribution of water as a resource is qualitative. The mineral, biological, and physical attributes of the water as it is available at specific locations may limit the use of the water for beneficial uses. For example, brackish water may be unfit for human and animal consumption and may have limited value for other purposes. One role for the field of water resource systems is to undertake the studies and research needed to enable sufficient water to be available within a specific locality or region in order to meet both present as well as future demands. Traditionally, it has been the task of the water resource professionals to apply techniques of systems analysis to this type of maldistribution problem and to identify ways to overcome the maldistribution of water and to mitigate, correct, and provide the resource on an as needed basis. This approach is still valid; however, it needs to be broadened to consider ways to alter the demand side of the water need equation as well as meeting the supply side of the water need equation.

Systems Analysis Methods: Multi-objective Planning

Normally one may expect a series of inter-related steps to be undertaken to accomplish a systems analysis of a problem of interest. Clearly, the methods of systems analysis may be applied to a wide spectrum of problems including but not limited to water resource problems and issues. The first step is to identify and quantify the objectives associated with the problem. For example, one objective may be to provide water of a specified quality and quantity to a specific region with a specified degree of reliability. A second related objective might be to provide this desired level of water service at minimum economic cost. Any problem that contains two or more objectives is considered a multi-objective problem. The solutions to this class of problem result in the identification of a range of potential solutions where the gain of one objective, for example, lower economic cost is achieved only through the tradeoff against other objectives — i.e., less reliable water service in the example. Included in the probl em definition portion of the systems analysis is the definition of decision variables, for example, the location and the level of treatment provided to achieve desired levels of water quality for the service area. In addition, the analysis requires the formation of constraints that describe the physical limitations associated with the problem at hand. The objective functions and the constraint equations which include the decision variables for the problem provide the analytical framework for the systems analysis to be undertaken.

A second step in the overall process is to collect data relevant to the problem. This data may be physical data, biological data, and/or chemical data that characterize the water resource in its present condition. Economic data including cost data is obtained which specifies information for this example on the costs associated with providing the desired level and reliability of water service. Simulation models and other techniques also may be utilized in order to calculate values of coefficients associated with the decision variables for the multi-objective optimization problem. Having assembled the necessary data and identified the objective functions and associated constraint equations, the third step is to generate alternatives to achieve the desired objectives. In carrying out this third step in the overall systems analysis, one is identifying potential solutions in objective function space, which satisfy all of the constraints in the problem and identify optimum solution points for the objective functions in the problem. For example, if cost is to be minimized one may set different levels of water service to be provided and then find the least cost solution to provide that specified level of water service.

The fourth step is to evaluate trade-offs between the objective functions that are being considered in the problem at hand. It is very important for the decision makers to have this information before them so that it is clear what increased risk they may incur if they choose to implement lower cost solutions for the provision of water service. It is this explicit trade-off between objectives that is a critical and important element in multi-objective planning for water resource development and use. This fourth step leads directly to the fifth step, namely the selection by the decision-makers of a preferred alternative from among the large number of potential solutions to the problem.

The sixth and final step is the implementation and continued maintenance of the preferred and selected alternative. This step may be the most complex of all of the steps in the systems analysis process. The implementation step requires the commitment of sufficient financial resources to build and to operate the water project. The ability to secure and maintain these financial resources often requires a major political commitment from the multiple units of government served by the project. It should be clear from this brief presentation that the planning and management of water resources is a demanding task requiring the contribution of knowledge and professional skills of many disciplines including planners, engineers, policy makers, economists and others.

Major Planning Issues: Resolution Needed

In undertaking a multi-objective planning activity in the field of water resources, there are a number of major issues which need to be addressed and resolved in order for the planning to proceed in an effective way. First of these issues is to decide upon the desired scale of development For example, is this water project to serve a limited locale or is it to serve a whole region or a whole watershed. Closely related to the scale of development is the sizing of the physical structural elements to be included in the project. This issue relates not only to the scale of development but also to the desired reliability to be achieved once the project is completed. For example, if the project is designed to protect human life, the reliability or safety factor associated with the project may require larger structures to provide the extra degree of protection of human life. If the water project involves certain control structures such as reservoirs then another issue that must be addressed is to develop a desired o perating policy for these control structures so that they can perform their function with a high probability of success. A further dimension to be considered in the planning of major water projects is to minimize undesirable environmental impacts associated with the implementation and operation of the projects. Finally, it is essential to provide for the establishment of an adequate organizational structure to implement and maintain the water resource project. This organizational structure needs to have the capacity to ensure the physical, financial, operational, technical, political, social, and environmental integrity of the project. This issue is among the most complex and challenging of all the tasks facing the water resource planners.

There are three major important issue areas that need to be considered to gain perspectives on global water problems facing the world as we move into the 21st century. These areas will be discussed and data will be presented to illustrate the impacts of these problems in certain areas of the world. The first key issue is the growth of global population against a relatively fixed supply of fresh water. The hard fact is simply that population growth is a driver that acts to increase the use of water to meet basic human needs. In addition to meeting the immediate individual need for water to sustain the individual, vast quantities of water are consumed through irrigation for the production of food and fiber to meet human needs. Additional large quantities of water are utilized for cooling purposes in the generation of fossil and nuclear energy. Given that the fresh water resources are not distributed uniformly across the globe, the increasing demand for the essential water resources requires that the water be u sed as efficiently as possible. Furthermore, it requires that means need to be found that will ensure the reliable provision of adequate food/fiber for those people living in water scarce regions.

A second key issue is the uncertainty associated with climate change and its potential impact upon the availability of fresh water resources in specific regions of the world. The risk of reducing the quantity of fresh water is of particular concern in regions that are already short of water needed to meet present needs. There can be increased evaporation as a consequence of global warming, soil moisture may decrease because of increased evaporation, and snowfall/snowmelt patterns may change. It is clear that the future will not be like the past; however, it is not clear what the future will bring other than change.

The third major global issue is that of sustainability and water resources. This topic is complex and not fully defined. The issue of population growth as a driver has already been discussed; it is important to attach a second driver to population growth and this second element is the standard of living. As the standard of living increases, increased demands are placed upon regional water resources to support the demands of people to achieve higher and higher standards of living. There are various technical means that may be utilized to augment existing water supplies. Desalination is very energy intensive and is applied in very special circumstances. Large scale, long distance transport of water has been practiced for many years to capture and move water from relatively water rich areas to areas where the demand exceeds the available supply. Diversion of water has increasingly come into question because once the water is diverted, it is highly unlikely that the receiving area will ever relinquish these augm ented waters. There are a host of problems that emerge as one considers sustainability and water resources. These include water quality and health, ground water management and use, environmental protection of source waters, transport and flood protection, natural disaster problems that include the loss of essential water services when areas experience hurricanes or earthquakes or sea level rise.

The problem of protecting against low probability events with severe consequences must receive additional attention. These elements of sustainability and water resources all have a common element, namely anticipation of change. Changes occur in water systems as the physical infrastructure ages. Changes are occurring in both the demand and supply of water resources. These changes will certainly continue into the future. It is very important that our water systems now and in the future incorporate the ability to adjust and remain viable in the face of unknown future stresses and changes.

There have been more modest increases in the percentage of urban populations now served for water supply and sanitation. While the absolute numbers of rural population not served by water or sanitation have decreased during this decade over the developing regions, the absolute numbers of population in urban areas not served by water supply and sanitation have actually increased. This outcome reflects the population driver as rural people leave the countryside to move to urban areas. The data for Africa is particularly important. It shows that the absolute numbers of unserved peoples in both urban and rural areas have increased over the past decade. In this region, the population driver has outpaced the provision of water and sanitation services. In the Middle East, the data shows that more progress has been achieved in the urban areas than in the rural areas.

It is important to note that Libya is withdrawing water at over 400% of its annual renewable water resources. This is clearly not sustainable; it is mining water from ground water under the desert and pumping it to the coast for human use needs. Examine the per capita withdrawn (cubic meter/person) in comparison with the overall world average.

Egypt, Sudan, and China all use 87% or more of their water withdrawn for agricultural purposes. Here the population driver coupled with the relatively fixed fresh water supply is clearly evident. The data shows a range of 30 cubic meters per capita per year (Egypt) to 9,940 cubic meters per capita per year for the United States. In the case of Egypt, the country is entirely dependent upon the inflow of the Nile River from the Sudan. The countries in the Middle East all have in common a very limited annual internal renewable water resource. Water is the life-blood for these countries and effective shared management of this resource is essential for sustainability. Increasing salinity of irrigated croplands, results from the high evaporation of irrigation water applied in these areas. The salts are left behind as the irrigation water evaporates. As the salinity increases, the soil becomes less productive and it requires crops with a high tolerance for salinity.

Observations

There are a number of important observations that we may draw from the information presented in this chapter. These observations include the following:

- It is imperative that we seek and implement the most effective and efficient use of water to meet human and environmental needs.
- Additional resources need to be committed to provide water to meet basic human requirements for health and survival.
- There is an important need to address issues of equity in terms of the use of limited water resources.
- It is important to recognize the uncertainty that exists regarding the potential adverse impacts of global climate change on fresh water resources.
- It is very important to educate the public as to the value and worth of water in society.
- There is a great need to utilize the watershed approach on an appropriate scale to ensure the most effective planning and management of water resources in the watershed.
- It is critical to provide the most effective and innovative institutional arrangements possible to ensure effective implementation and sustained operation of key water resource facilities.

In autumn 2002, thousands of decomposing chinook salmon, a threatened species under the Endangered Species Act, (1) lined the dry river bed of the Klamath River and permeated the air with the smell of unnecessary death. It was the worst fish kill in American history, with 34,000 to 70,000 adult fish carcasses lining the banks of the Klamath River for thirty miles. Although these salmon perished due to two fish pathogens, a contributing factor was a shortage of water in the Klamath River. In the last five years, virtually every region of the United States has experienced a water shortage and, by 2013, at least thirty-six states anticipate some sort of water shortage. Shortages occur because virtually every aspect of American life ties itself to water, with vast amounts devoted to producing electricity, growing food, manufacturing household goods, and serving other personal uses.

Although water shortages have long occurred in the United States, evidence suggests these shortages will continue to worsen, especially in the arid West. Inadequate implementation of the doctrine of prior appropriation, the most popular water system in the western United States, exacerbates this continued decline. First, the doctrine gives priority to the earliest water users; thus, if a shortage occurs, the state dispenses water in the order it granted permits and cuts off the most recent users. Second, a water user may divert and use as much water as the diverter can put to "beneficial use." Thus, under the prior appropriation system, the state grants each user a certain allocation of water which the user may not exceed, and which must be put to beneficial use. However, because states largely have failed to measure the amount of water a user diverts, a process known as "source metering," it is unclear whether users are complying with the conditions of their state permits. In 1993, Washington became the first western state to require the measurement of virtually all surface water withdrawals, allowing the state to determine the amount of water diverted from rivers and effectively manage water supplies. The Washington legislature created this source-metering law, part of a larger package aimed at promoting salmon recovery, to ensure compliance with water appropriation permits, protect instream uses, and help determine whether the state has water available for appropriation. After fifteen years and multiple revisions, the statute appears to have made the state's management of water more efficient. However, no other western state has followed in Washington's footsteps.

In light of climate change, all western states should adopt source metering. One study suggests that in the West, "no other effect of climate disruption is as significant as how it endangers... already scarce... water supplies." Climate change influences water supplies mainly by affecting precipitation. As temperatures rise, less snow falls in the West and snowpacks shrink, making less water available. In Washington, on the other hand, source metering has helped to leave more than 300,000 acre feet of water in streams. Because source metering leaves more water in streams, it may be the most effective tool to ensure the efficient functioning of the prior appropriation doctrine in a climate-changing world. Kansas and Texas have implemented some form of source metering, mainly to deal with water shortages. Additionally, WaterWatch of Oregon, a fiver conservation group in the western state of Oregon, proposed a state bill to require source metering throughout the state, but the bill died in the 2007 state legislative session. Curiously, the western states, notoriously plagued by water shortages, seem the most resistant to source metering laws, even though they have the most to gain from efficient management of water resources.

Water Consumption and Waste

In the last fifty years, world water consumption has tripled. Water use in the western United States continues to rise as the demand for water to provide energy and support agricultural and metropolitan uses, recreation, fish and wildlife habitat, and water quality protection increases. The climate of the West, with its arid, desert-like lands that receive less than twenty inches of rainfall per year, exacerbates water shortage problems. However, even in the wet areas of western Washington and Oregon, where rainfall can exceed more than 100 inches per year, water shortages have become a concern.

Figure: *According to the World Water Council, global fresh water consumption has increased six times over during the twentieth century, accompanying a tripling of the population in the same period.*

Agricultural Use

An estimated 408 billion gallons of water were withdrawn in the United States for all uses in 2000. Of this amount, more fresh water is used in agriculture than for any other use. Irrigation uses the overwhelming majority of water consumed in western states. For example, irrigation withdrawals consume 80% of all water used in Utah and 90% of all the water used in New Mexico. Additionally, many crops grown in the West are low-value crops. In California, pasture, alfalfa, cotton and rice—the four largest water-using crops—use over 50% of all agricultural water. However, the economic value of all these crops together is similar to that of the state's grape crop, which uses only one-ninth as much water. History is one reason for such inefficiency: states granted very generous water rights to early farmers, especially those raising livestock, and the farmers continue to pass down these property rights in water through generations. Existing water users have essentially fully appropriated all available water, meaning that new users can only obtain water when prior existing uses change. Although agriculture traditionally employed more than half of the western population, this number has been on the decline. By 1991, the natural resource industries together provided less than 6% of employment in the West. As agriculture employs fewer people, and the demand for water remains high, water use has shifted away from agricultural use and toward economically higher-valued industrial and municipal uses. Farmers have begun to market their water rights, realizing they can make more money selling their water supply than by growing low-value, water-intensive crops like alfalfa and rice.

Urban Use

Population in the West has exploded over the last few decades. Spatial changes have been the most significant, with people moving away from rural areas and congregating in urban areas. As western metropolitan areas grow, so does the demand for water. In fact, the most recent report from the United States Geological Survey found that municipal withdrawals increased by 8% between 1995 and 2000 alone. Homes account for more than half of the municipal withdrawals, representing much greater consumption than either business or industry. Location also causes these amounts to increase. For example, the arid West has very high per capita residential water use, due to landscape irrigation. As more water moves toward urban use, the use becomes less elastic because a municipality must always provide water for the basic needs of its citizens. On the other hand, a farmer can forgo applying water to his crops during a year of shortage. Projections suggest that people will continue moving to the West for at least the next twenty-five years, meaning the demand on water will remain high and become less elastics.

Salmon

In Washington and throughout the Pacific Northwest, salmon have played a critical role in history, culture, economy, and recreation. Tribal people value salmon for subsistence and their cultural significance, fishermen value them for sport and economic importance, and environmentalists value them for their ecological significance. Habitat loss, however, poses a considerable threat to the existence of salmon. Wild salmon declined drastically during the twentieth century; by 1999, salmon had disappeared from 40% of their historic spawning grounds in Washington, Oregon, California, and Idaho. Water shortages often cause habitat loss because salmon need clean, cool water in order to survive. Low streamflows can interfere with upstream migration and may reduce or even eliminate spawning habitats. In 1991, the federal government placed one population of Washington salmon on the list of Endangered Species List, partially due to lack of habitat. Over the next eight years, the federal government listed three additional populations of salmon in Washington as endangered or threatened. Lack of adequate salmon habitat played a dominant role in spurting the government to create these listings.

Recognizing that salmon populations would not thrive without drastic changes, the Washington legislature enacted a legislative package to address salmon recovery. One of the new laws established a requirement that all new and certain existing surface water users within the state measure their surface water diversions. In response to these requirements, Washington became the first western state to implement source-metering at the statewide level. The slow implementation of Washington's 1993 source metering statute prompted a lawsuit in 1999 to spur completion. As a result, the Washington State Department of Ecology (WDOE) was ordered to submit a compliance plan. Thus, fifteen years after its enactment and several lawsuits later, the statute finally seems to have achieved its goal of efficient state water management.

The Beginning of the Source Metering Statute

The 1993 law required all new and certain existing water users to measure surface water diversions as part of a larger salmon recovery package. The law required measurement of every 1) new surface water permit, 2) existing surface water permit exceeding one cubic foot per second (cfs), and 3) all new and existing permits, regardless of size, in areas where salmon stocks are "depressed or in critical condition." Essentially, the only water users not regulated by the new statute consisted of those existing surface water users with permits of less than one cfs outside of critical salmon areas. The legislature intended this statute to be the first step toward effective water management. According to the legislature, water management requires information gathering, meaning the state must know who is using what amount of water and when they are using it. Without this information, the WDOE could not efficiently identify any illegal uses of water, nor could it provide adequate water for fish, resolve conflicts between water uses, or promote conservation.

Although state law required source metering in 1993, the WDOE failed to adopt implementing regulations. Without new regulations, the existing regulations did not require, or even allow, metering of any kind. The WDOE also failed to apply the source metering rule to groundwater. Although the original statute applied only to surface water, the statutory provisions regulating groundwater incorporated all surface water regulations into the groundwater code. Thus, when the legislature required source metering for surface water, the rules should have applied to groundwater as well.

In 1999, as a response to these deficiencies, a coalition of environmental groups filed suit, contending that the WDOE failed to implement administrative rules regarding water diversions, particularly implementing the requirement to establish water metering. The Thurston County Superior Court agreed, ordering the WDOE to create a compliance plan that included adopting either a new or revised administrative rule requiring source metering by December 31, 2001. In addition, the court ordered the WDOE to require the metering of 80% of water use in each of the sixteen critical fish basins by December 31, 2002.

The WDOE responded by creating a compliance plan and adopting a revised administrative rule on source metering by the required deadline. By

2003, the WDOE issued administrative orders requiring metering for 903 water rights, which amounted to 80% of the total estimated water diversions in the critical fish basins. In order to ensure that water users met these requirements, the state legislature appropriated over $3 million to defray the installation costs of the metering equipment. Moreover, the WDOE's rules required measuring devices on all new water rights for both surface and groundwater withdrawals and source metering for all water users requesting a change or extension to an old right.

The Current Source Metering Statute

Washington's source metering statute gave the WDOE authority to require that all new and certain existing water users measure their water diversions. The WDOE rules established standards for measuring devices as well as for recording and reporting water use data. The agency's goal was to make sound water allocation decisions based on "the reliable, accurate measurement of state water that is diverted, withdrawn, stored and used." The WDOE also aimed to determine the availability of water for apportionment among new users, while enforcing water rights compliance and protecting instream flows.

The rules require source metering by anyone seeking a new surface water permit and by any existing permittee diverting in excess of one cfs. The rules also apply to all permits in areas where salmonid stock are either "depressed" or in "critical" condition. However, they do not apply to secondary users, such as customers of public water supplies and members of public irrigation districts, because that would result in duplicative and unnecessary expense. Any water user subject to these rules may be required to inform the WDOE about its diversions, including the location, the flow rate of diverted water, and the type of measuring device.

The WDOE rules require recording monthly measurements for diversion rates less than ten gallons per minute (gpm), biweekly measurements for diversions from ten to forty-nine gpm, and weekly measurements for diversions over fifty gpm. The measuring device must generally be installed and maintained to the manufacturer's specifications and be in good working order. If a water user fails to comply with these requirements, the WDOE has authority to levy civil penalties ranging from $100 to $5000 per day.

Although a water user bears the primary responsibility of complying with the source metering statute, the state legislature provided money to defray some of those costs. To qualify for cost-share assistance, an applicant must have a valid water right and use the money only for metering projects. Availability of costshare assistance begins only after the first $250, with a maximum state share of $45,000. Water users in fish-critical basins have priority for assistance.

After the 1993 adoption of the source metering statute and the subsequent legislative revisions because of the 1999 lawsuit, the majority of the state's water users now comply with the statute. This widespread compliance provides WDOE with measurement information to ensure compliance and promote conservation. (112) The source metering statute, coupled with regional recovery plans and international treaties to protect salmon, has led to more than 300,000 acre feet of water left in streams where salmon need it most, demonstrating the statute's success.

In the western United States, only Washington has implemented source metering at a statewide level. However, both Kansas and Texas each require some form of source metering. Certain regions in Texas require groundwater users to measure diversions. and Kansas requires water use reporting for particular water rights holders. Environmental groups in Oregon, such as WaterWatch of Oregon, have attempted to pass statewide source metering laws, but so far have been unsuccessful.

Source Metering in Texas

Population growth in Texas exceeds every other state in the nation. Studies suggest Texas will double in population by 2050. As a result, the state's current dependable water supply will meet only 70% of the projected water demand by 2050. Increased demand for water due to the growing population has led certain Texas water districts to encourage and even require specific users to install water meters, but the legislature has not yet taken any statewide action. Texas divides itself into five locally owned, landowner-operated, soil and groundwater conservation districts. Texas is also partitioned into numerous water improvement districts that focus on surface water irrigation, mostly corresponding to county lines.

Many conservation and improvement districts have created rules imposing source metering, all varying from one another. Some districts differentiate among domestic, industrial, and agricultural water users to determine who must measure water consumption. For example, select districts require metering on wells capable of producing more than 25,000 gallons of water per day, essentially exempting domestic users. Also, some districts require water users to employ only approved manufacturers and models of water meters, while other districts require that water meters "meet the American Water Works Association's accuracy reading range for actual flow." Certain water managers conduct random checks to verify accuracy; others require certification tests or request third party tests of the meter. Virtually all districts that require source metering expect the water user to keep track of the readings and provide information about the actual amounts pumped to the water manager. For the most part, the conservation districts provide no money to water users to defray the cost of the meters, although one surface water irrigation district does have a 50% cost-share program.

Although the Texas legislature has yet to implement source metering statewide, support for statewide source metering does exist. In 1997, the Texas legislature passed legislation addressing future water demands in the state. The legislation tasked regional water planning groups with developing water demand projections, and strategies to meet those demands, through 2050. The Texas Water Development Board then compiled these projections and strategies into a state water plan. Twelve of the sixteen planning groups recommended "water use management, such as... volumetric measurement of water use." Although the suggestion applied only to irrigators, it reflects widespread support for source metering in Texas.

In comparison to Washington, Texas has not mandated source metering throughout the entire state. While only some water districts in Texas have mandated source metering, all of the districts have implemented varying rules regarding source metering, making it less streamlined than Washington's system. Almost none of the Texas districts provide monetary assistance to defray the cost of compliance, whereas Washington provides financial assistance after the first $250 spent. However, both Washington and Texas require that water users who must meter provide the state with the source metering data to demonstrate compliance.

Source Metering in Kansas

Kansas farmers produce more wheat and sorghum than any other state in the nation, making agriculture the largest industry in the state. Cattle production also thrives throughout the state. These agricultural pursuits require considerable amounts of water, something Kansas, especially its western half, has little to spare. Although Kansas may not have the population crush affecting the western United States, water shortages still occur due to the high demand for water in the agricultural sector. The majority of water in Kansas comes from the Ogallala Aquifer, which underlies most of the midwestern United States.

Only a finite amount of water exists in the Ogallala Aquifer, but water users—especially irrigators—are rapidly depleting these water levels. As a result of agriculture's high demand for water in Kansas, the state has long recognized the importance of water measuring efforts. Beginning in 1980, Kansas groundwater management districts began to require meter installation for new and redrilled wells. Ten years after the districts initially implemented metering requirements, the Kansas legislature enacted legislation requiring annual water use reporting. Two rules govern source metering as a mechanism for reporting water use in Kansas.

The first gives the Chief Engineer of the Division of Water Resources of the Kansas State Board of Agriculture authority to require any water user to install a measuring device. An agent of the Chief Engineer may read the measuring device "at any time" and may require users to report "at reasonable intervals." The agent also has authority to require a water user to report findings of water waste or water quality issues. The Division of Water Resources has enforced this authority by requiring measuring devices on all new or changed points of diversion since 1987. Kansas also requires measuring on existing points of diversion where 1) water fight administration regularly occurs, 2) the state needs better data to comply with an interstate fiver compact, 3) the state has created intensive groundwater use control areas, and 4) the state administers "minimum desirable streamflows." The second rule governing source metering in Kansas gives authority to every groundwater management district to install measuring devices on water users' points of diversion or to require that water users install the measuring devices themselves. The district may either read the measuring devices to determine water flow or require a water user to read their own device and report the readings "as may be necessary to determine the quantity of water withdrawn." Every groundwater management district requires water meters for almost all nondomestic groundwater pumping. The districts also have authority to give assistance to water users in their district. This authority includes providing help with testing and maintaining water meters, as well as filling out water use reports and water appropriation applications. In total, over 30,000 of the approximately 38,000 active points of diversion in Kansas require source metering. According to one account, Kansas has created the "best water use reporting program in the United States." The Kansas Division of Water Resources uses eight employees to implement this apparently successful program. These employees analyse the annual water use reports from metering water users to review and assemble a statewide annual water use report. Participation among water users in annual water user reporting remains high: each year 99.9% of all water users file a report. There is a civil fine of up to $250 per water right for failure to file an accurate and complete report, which may encourage compliance.

Both Washington and Kansas require statewide source metering, although the states have different classifications as to who must meter. Also, Kansas requires metering for all new or changed points of diversion post-1987, whereas Washington state law requires source metering for all new and some existing users post-1993. Both Kansas and Washington require water users to file reports in order to demonstrate compliance with the state laws. Kansas and Washington require measuring of both surface and groundwater, although Washington regulates both under the same statutory provisions and Kansas divides the regulation between the Division of Water Resources and the groundwater management districts. Each state has noncompliance measures in place to penalize those who fail to meter.

Source Metering in Oregon

Oregon has a national reputation for its progressive environmental protection efforts. The state imposes progressive land use planning restrictions on development and considers instream river flows to be a state water right. Although Oregon often pioneers environmental change, the state has yet to enact a statewide source metering rule. WaterWatch, a fiver conservation group dedicated to protecting natural flows in Oregon's rivers, has made several attempts to encourage statewide source metering.

Figure: *Oregon is famous for its coastline protection efforts*

Until recently, these efforts failed to gain political traction. However, in the 2007 legislative session, Democratic control of the House, Senate, and Governor's office resulted in a new political dynamic, allowing a variety of interest groups to move forward on issues which had garnered less attention in the past. Thus, when WaterWatch proposed a bill calling for statewide source metering in 2007, the state legislature took notice. The original House bill called for all water users to measure and report uses, under the notion that "what gets measured gets managed." Although current state "law requires certain municipalities and irrigation districts to measure and report water use, this bill attempted to codify a requirement that all unmeasured water fight holders must measure diversions. The bill would have implemented a cost-sharing program to help defray the cost of purchasing the measurement devices.

The proposed bill created considerable controversy. Although the bill had the support of many influential groups, the agricultural community vehemently opposed it. The agricultural community feared the cost of source metering and contended that the passage of the bill could lead to more intrusive laws that would infringe upon their water rights. With no consensus, a work group convened by the chairperson of the House Committee on Energy and the Environment recommended a compromise version of the bill that called for the measurement of significant diversions in priority watersheds and provided for a cost-share program. Although the bill died before this committee, WaterWatch moved the source metering concept further along than in any prior session and claims to have "built a strong coalition around the bill."

The source metering directives in Washington, Texas, and Kansas can serve as examples for Oregon as it pushes forward with the goal of statewide source metering. Oregon may use Washington's model to show how a western state has benefitted from source metering, while also serving as a cautionary tale that political strife and lawsuits may occur before implementation. Texas provides an example of district-wide source metering, a process with as many different rules as districts. Texas's use of a less streamlined approach with little uniformity and no financial assistance may work for a district wide approach, but Oregon may find such a plan difficult to implement at a statewide level. Finally, Kansas provides Oregon with a more established model for statewide source metering that demonstrates the benefits of statewide source metering.

The Benefits of Source Metering

Although water users may have concerns about source metering, primarily due to cost and concern for increased water right restrictions, the benefits clearly outweigh these concerns. The water itself costs nothing; however, a water user typically incurs some initial cost for purchasing and maintaining the source meter, although the source metering usually pays for itself over the long term.

Additionally, source metering does not automatically lead to increased water use restrictions—it has not in Washington, Texas, or Kansas; it only provides accountability for water use in the system. Source metering may also give water users an effective mechanism for responding to climate change by providing an effective mechanism for managing a state's scarce water supply.

Concerns Over Source Metering

Water users frequently have two primary worries over requiring source metering: the cost and the potential for increased water use restrictions. Water users do bear some of the cost of implementing.metering, which typically costs from $600 to $2500 depending on the size of the diversion and type of meter involved. However, a measuring device can function for up to twenty years. Yearly maintenance generally costs about $200 per meter. A water user pays nothing for the water it diverts. Although water users bear some costs,

Washington provides a cost-share program that helps to defray some of these costs. Source metering, in fact, often pays for itself over the long term because it tends to produce water savings, lower energy costs, and optimum yields.

In Texas, some farmers found that source metering immediately paid for irrigation systems tied to multiple wells. Another Texas water user saved almost $2000 after discovering that although the pumps on a water well were running, no water flowed from the well. Yet another Texas study concluded that water measurement alone reduced water usage by 10%. An Oregon study showed that water measurement alone can decrease diversions and increase streamflows; when the state ordered the measurement of diversions on the upper Wood River in the Klamath River Basin, fiver flows increased by twenty to thirty cfs. Perhaps most significantly, in Washington, more than 300,000 acre feet of water has been left in streams where salmon need it most, protecting aquatic species against habitat loss and perhaps creating a buffer against climate change. There is no evidence that source metering leads to an increase of water use restrictions.

Under most state water laws, the state specifies the amount of water a water user may divert; thus, water metering cannot take away the water right of any water user. Instead, source metering helps only to ensure that all water users fall within the scope of their water rights. Source metering allows the state to gather information, but does not necessarily cause the onset of regulation. No examples of source metering leading to an increase in water use restrictions in Kansas, Texas, or Washington could be found.

Climate Change

Global surface temperatures have increased by 1.1 degrees Fahrenheit during the past century. Average temperatures in the West have risen higher than in any other region in the contiguous United States, increasing by two to five degrees Fahrenheit during the twentieth century. Climate change influences water supplies mainly by affecting precipitation. As the temperature increases, snowfall and snowpacks have decreased throughout the West, reducing the amount of water available for reservoir storage. Snowpacks melt earlier due to higher temperatures, requiring more reservoir storage to capture ‘and hold the runoff until summer, the highest demand period for water.

Studies have already documented the occurrence of climate change in the West. According to National Weather Service data, the western United States, along with the rest of the world, has warmed. With warmer temperatures, less snow has fallen and western snowpacks have reduced in size. One study of 200 western mountain sites discovered more than two-thirds of the sites showed less winter precipitation falling as snow and more as rain. Another study analysed the records from 824 government snowpack measurement sites in the West and found that snowpack levels had declined at virtually all of the sites from 1950 to 1977. Yet another study found that western snowpacks are melting earlier in the year; this study looked at 279 western rivers and streams dominated by snowmelt to determine that the timing of peak flows advanced from ten to thirty days from 1948 to 2000.

Although these studies depict changes that have already occurred, further studies suggest these changes will only continue in the future. For example, a growing body of scientific research indicates that many parts of the western United States will experience reduced water availability in the future, especially during the high-demand period of the summer months. In fact, one study proposes that in the West, "no other effect of climate disruption is as significant as how it endangers the region's already scarce.., water supply." Climate change has already occurred in the West and will likely only continue into the future.

Source metering can mitigate the effects of climate change by providing an effective mechanism for managing a state's scarce water supply. Source metering provides a state with information about the amount of water a water user consumes and the amount of water left in a stream, making it easier for a state to manage the water supply. Source metering has been shown to produce watering savings, optimum yields, and lower energy costs and has reduced water usage in Texas, Kansas, and Washington. As climate change produces more water shortages, the additional water that source metering leaves in a stream may mitigate the effects of rising temperatures by providing the extra water a state needs both for its water users and for protecting instream flows.

Western states, which are experiencing a disproportionate amount of climate change-induced water shortages, should follow Washington's statutory model and require virtually all water users to measure both surface and ground water diversions. These states should couple a source metering requirement with a cost-sharing program and perhaps a loan program, so that water users do not bear the full cost of the program, because the entire state would ultimately benefit from this program. Western states should also realize that implementation of source metering may be difficult: the regulating body may not implement the needed rules immediately and lawsuits may be filed, as in Washington, and senior agricultural water users may be resistant to the passage of such a statute, as in Oregon. However, source metering may be the most effective mechanism available to ensure the efficient functioning of the prior appropriation doctrine in a climate-changing world in which there is surely going to be less available water in the future.

Washington created the first western statewide source metering statute to promote salmon recovery, determine the amount of water diverted from rivers, and effectively manage water supplies. Washington's source metering statute has led to effective management of the state's water supply by providing the state with needed information about the water supply and by leaving more water in the streams.

Additionally, Texas and Kansas provide examples of successful source metering statutes in nonwestern states, while Oregon exemplifies the struggle of adopting source metering in a western state. Although these examples strengthen the notion that Washington has provided a successful model for statewide source metering, no other western state has followed suit. Due to the efficacy of source metering and the impending consequences of global warming, all western states should adopt a source metering statute similar to Washington's. This statute provides a successful model for use in western states in light of the arid nature of the West, the increase of population, the decrease of water availability in this area, and the onset of climate change. To ensure the full value of water in the West, states must move toward counting every drop.

2

The Water Quality Trilogy

The opening provisions of many federal environmental laws contain broad, laudable goals. Saying and doing, however, are two entirely different matters. In the 1972 amendments to the Federal Water Pollution Control Act, now known as the Clean Water Act (CWA or the Act), Congress articulated one of the broadest ecosystem restoration and protection aspirations in all of environmental law: "The objective of this chapter is to restore and maintain the chemical, physical, and biological integrity of the Nation's waters." Courts have referred to this broadly stated objective as the "guiding star" of the statute, or otherwise invoked the language as the main starting point for their analysis.

But while progress has been made in moving toward "chemical" integrity, and while significant resources and programs have been directed at discharges of chemical pollutants, both the "physical" and "biological" integrity books in the trilogy have remained largely hortatory Empirical evidence shows measurable gains in reducing chemical pollution, but in the thirty years since the law was passed, the overall health of the nation's freshwater aquatic ecosystems has declined dramatically. Disturbing and ongoing trends in the health of our aquatic ecosystems, relative to the lofty objective announced by Congress three decades ago, suggest several threshold questions First, did Congress understand the implications of its tri-part "integrity" objective? It is possible, at least, that in 1972 Congress believed—erroneously but perhaps understandably—that "chemical, physical, and biological integrity" could be "restored and maintained" largely by stemming the flow of chemical pollutants from various sources.

Perhaps Congress did not understand fully that by invoking physical and biological as well as chemical factors, it was implicating a far broader range of human insults to aquatic ecosystems, such as dams and diversions, channels and levies, bridges and culverts, and waterside construction of roads, homes, resorts, and other structures. Second, if Congress did understand those implications of the Act's threshold objective, did it really mean what it said?

It is all too easy for Congress to purport to address environmental problems in aspirational declarations, but either to leave the hard and important choices and details of implementation to the Environmental Protection Agency (EPA) and other federal and state entities, or to omit adequate implementation strategies altogether.

Perhaps Congress intended to identify physical and biological integrity as desirable, if not essential, objectives; but recognizing the significant political and implementation barriers to achieving those objectives, intentionally failed to match them with effective implementation tools. Third, if Congress both understood the implications of the Act's overarching objective and actually meant what it said, why have those objectives not been met? Are the statutory tools provided in the Act inadequate to the task, or have they been implemented poorly? Is EPA largely responsible for the failure, or do the states share a large portion of the blame?

If the problem is the latter, should more authority in the realm of physical and biological integrity be shifted from the states to the federal government? In short, what can we do better or differently to reverse persistent declining trends in the health of aquatic ecosystems, and to move toward the goal of chemical, physical, and biological integrity of the nation's waters? This analysis finds that Congress did mean what it said in establishing the broad, tri-part aquatic ecosystem protection objective of the CWA. However, neither the federal nor the state agencies charged with implementing the CWA have taken full advantage of their existing legal authority to address the physical and biological books in the water quality trilogy.

Of course, in the 1972 Act and later amendments, Congress did provide far more precise tools to address the chemical as opposed to the physical and biological integrity aspects of the water quality trilogy, entrusting authorship of the latter two books largely to the states. As a result of both factors—inadequate statutory authority and inadequate implementation of the authority that does exist—the health of the nation's aquatic ecosystems has continued to decline precipitously over the past three decades. It is time for EPA to revisit its virtually exclusive focus on chemical impairments to our aquatic ecosystems, and for Congress to revisit its policy of deference to the states in areas other than point source control of chemical pollutants.

Defining the Water Quality Trilogy

Congress divided its CWA aspirations into a broadly defined, overarching objective, followed by a series of subsidiary or, sometimes, interim goals. In section 101(a), Congress announced boldly that "[t]he objective of this chapter is to restore and maintain the chemical, physical, and biological integrity of the Nation's waters." "In order to achieve this objective," Congress identified a list of seven goals, one of which is particularly relevant to this chapter: "[I]t is the national goal that wherever attainable, an interim goal of water quality which provides for the protection and propagation of fish, shellfish, and wildlife and provides for recreation in and on the water be achieved by July 1, 1983...."

It appears this "interim" water quality goal, often referred to by the misleadingly narrow shorthand "fishable and swimmable waters," has become the de facto ultimate water quality objective in lieu of the broader objective defined by Congress. An obvious threshold question is whether and how Congress intended the broader statutory objective to be effectuated relative to the Act's subsidiary goals.

One potential answer is found in section 303(c) of the Act, which establishes minimum requirements for states (or EPA) in adopting ambient water quality standards (WQS): [R]evised or new water quality standard(s) shall consist of the designated uses of the navigable waters... and the water quality criteria for such waters based upon such uses. Such standards shall be such as to protect the public health or welfare, enhance the quality of water and serve the purposes of this chapter. Such standards shall be established taking into consideration their use and value for public water supplies, propagation of fish and wildlife, recreational purposes, and agricultural, industrial, and other purposes, and also taking into consideration their use and value for navigation.

Clearly, the reference to "propagation of fish and wildlife" alludes to a key portion of the interim water quality goal set forth in section 101(a)(2). The requirement that WQS be adopted to "serve the purposes of the Act," however, goes further to require those standards to effectuate all of the statutory purposes, including both the subsidiary goals and the overarching objective. Thus, section 303(c) converts an advisory statement of legislative intent, which normally has no legal force and effect, into an operative provision of substantive law. However, EPA adopts this view only in part. Congress reinforced the distinction between the release of chemical pollutants and the broader notion of restoring aquatic ecosystem integrity by adopting different definitions of "pollution" and "pollutant" in the 1972 law.

It defined "pollution" as "the man-made or man-induced alteration of the chemical, physical, biological, and radiological integrity of water," language obviously intended to embrace the full agenda articulated in the overall statutory objective. The term "pollutant," by contrast, is defined more narrowly as specified categories of discrete substances, the discharge of which into water is either prohibited, allowed, or qualified according to various provisions of the Act. The pollution/pollutant distinction, therefore, is a linguistic surrogate for the breadth of issues addressed in the overall objective of the CWA compared to its more finely focused subsidiary goals Indeed, of the seven subsidiary goals set forth in section 101(a) of the Act, only the last, which discusses nonpoint source pollution, speaks in the broader language of pollution rather than pollutants. Again, the obvious threshold question is whether and how Congress intended to achieve the broader goal of pollution control as opposed to the more limited control of "pollutant" discharges.

The potential significance of the distinctions between pollution and pollutants, and between chemical as opposed to physical and biological integrity, is best appreciated by reflecting on the types of activities that might comprise each category of activity and its resulting harm. Setting aside for a moment the actual language of the CWA, the common understanding of the term "water pollution" typically involves images of chemical pollutants spewing out of factories and sewage treatment plants, along with the resulting odor, discoloration, and harm to fish, drinking water, and other downstream water uses. Fueled by contemporaneous reports of dangerous chemical contaminants in seafood and drinking water, fish kills due to chemical pollutants, and dramatic incidents such as the famous fire on the Cuyahoga River, perhaps it was such images that influenced to enact the CWA in 1972.

The statutory objective of the CWA and the definition of "pollution," however, suggest a far broader array of human activities not typically included in the lay understanding of water pollution. Both provisions apply the concepts of chemical, physical, and biological integrity to the words "the Nation's waters" and "water," respectively. Even this simple distinction within the statutory text suggests some potential ambiguity.

Construing the statute most narrowly, one could determine that Congress was referring to the integrity of water itself, as opposed to the much broader concept of the aquatic ecosystems as suggested by "the Nation's waters." On its face, however, it appears nonsensical to speak of the biological integrity of water, unless Congress meant simply to exclude chemical pollution of such intensity that living organisms cannot be supported. More important, the legislative history outlined below indicates irrefutably that both the House and Senate Committees that sponsored the 1972 Act intended the phrase to apply to aquatic ecosystems and not just the integrity of the water within those systems. Thus, the Supreme Court has recognized that Congress was concerned with the protection of both water quality and entire aquatic ecosystems.

If so, what human activities other than discharges of pollutants are encompassed within the definition of "pollution," which applies to any human alteration of the chemical, physical, and biological integrity of the nation's waters (i.e., the nation's aquatic ecosystems)? Consider the following:

1. Common physical modifications to aquatic systems, such as armoring of river banks, channelization, levies, culverts, and other structural changes to rivers and lakes, which are designed to prevent erosion and flooding, radically alter the physical integrity and continuity of the nation's waters.
2. Widespread human introduction of non-native species (e.g., the stocking of non-native rainbow trout in rivers and lakes or nonnative macroinvertebrates to augment the food supply for those populations) can radically alter the biological integrity of the nation's waters.
3. The impoundment of rivers into reservoirs by hundreds of thousands of U.S. dams, both large and small, transforms entire aquatic ecosystem types.
4. Substantial diminution of instream flows, which is virtually ubiquitous in the West and increasingly present in eastern systems as well, radically alters the hydrological integrity of the Nation's waters.

These examples underscore the importance of Congress's intent when Congress established the ambitious goal of restoring and maintaining the "chemical, physical, and biological integrity of the Nation's waters." Did Congress understand the full implications of the broad language it used, and if so, how did it intend those ambitions to be implemented? Obviously, this requires an inquiry into what Congress meant by the words "restore and maintain" and "integrity," as well as the different ways in which the statute addresses pollution and pollutants. Those questions, in turn, require a more detailed evaluation of the text and legislative history of the 1972 Act and relevant subsequent amendments.

Textual Evidence

The first approach to this issue is to explore the full text of the statute to identify additional evidence of how Congress intended the narrower issue of pollutants and the broader issues of pollution to be addressed. In particular, did Congress use the terms "pollutants" and "pollution" in a logical pattern, which would indicate a conscious design in the overall statutory regime as opposed to more casual or unintentional differences in language? Congress distinguished between discharges of pollutants and the broader concept of pollution in stating the subsidiary goals of the Act, using the term "pollutants" largely for goals involving discharges and "pollution" in reference to nonpoint sources. How Congress actually intended these distinctions to be addressed, however, requires a more detailed analysis of the operative provisions of the law Unfortunately, this exercise produces somewhat mixed results.

Definitions

First, having established different statutory definitions of pollutant (narrower) and pollution (broader), Congress also defined other statutory terms selectively, in a pattern that appears to make sense Some of the other statutory definitions refer specifically to pollution while others speak in the language of pollutants and discharges Sections 502(1) and (2) define "state water pollution control agency" and "interstate agency," respectively, as having responsibilities to control pollution. Sections 502 and 502, by contrast, which define the precise coverage and operation of the point source control provisions of the Act, use the narrower term "pollutants." This usage seems to make sense to the extent Congress, as explained below, intended the broader issues of pollution, with their greater implications for land and water use, to be addressed primarily by state and interstate agencies as part of their nonpoint source pollution control efforts, as opposed to EPA and other arms of the federal government. It is also consistent with Congress's use of the term "pollution" only in the subsidiary statutory goal aimed at controlling nonpoint sources. The contrasting definitions, therefore, again suggest Congress used the different terms intentionally, in ways that match the Act's overall allocations of responsibility in the statute.

Other uses of these terms, however, could suggest less precision and intentional design For example, in section 502, the term "effluent limitation" is defined as "any restriction... on quantities, rates, and concentrations of chemical, physical, biological, and other constituents which are discharged from point sources...." Initially, this definition appears to conflate the notion of chemical, physical, and biological integrity associated with the broader definition of pollution, with the narrower definition of pollutants associated with discharges from point sources. On the other hand, the statutory definition of pollutant includes examples of chemical, physical, and biological constituents. That explanation, however, begs an important question: Did Congress intend that chemical, physical, and biological integrity be limited to the discharge of chemical, physical, and biological pollutants; or, alternatively, did Congress simply recognize that the broader concept of chemical, physical, and biological integrity encompasses the discharge of these pollutants as well as other forms of human alteration?

A similar issue is suggested by the definition of the term "discharge," which, "when used without any qualification includes a discharge of a pollutant." This language suggests that the unqualified term "discharge" covers activities broader than those in the qualified phrase but does not expressly indicate how. In Oregon Natural Desert Association v. Dombeck, the Ninth Circuit rejected the logical argument that, for purposes of applying the state water quality certification provision in section 401 of the Act, the word "discharge" includes pollution from both point and nonpoint sources. Instead, the Ninth Circuit panel accepted the government's alternative theory that the term "discharge" "is limited to point sources but includes both polluting and nonpolluting releases," a distinction that appears to makes little sense, or at least serves no apparent purpose in the statutory scheme. Even the broader interpretation, of course, would encompass at most discharges from both point and nonpoint sources, as opposed to other types of alteration of aquatic ecosystem integrity.

Operative Provisions

More insight might be obtained by examining the operative provisions of the statute to determine whether use of the terms "pollution" and "pollutant," or any further reference to "chemical, physical, and biological integrity," suggest a logical and intentional distinction. Once again, however, such a pattern is reflected in some, but not all, of those provisions.

Some operative provisions of the CWA very clearly appear to acknowledge the precise distinction between pollutants and pollution, indicating Congress understood quite well the distinction between the narrower goal of reducing discharges of pollutants and the much broader goal of abating pollution to restore and maintain the chemical, physical, and biological integrity of the nation's waters. For example, section 105(c) provides for "contracts... for research and demonstration projects for prevention of pollution of any waters by industry including, but not limited to, the prevention, reduction, and elimination of the discharge of pollutants." This language suggests Congress understood, even with respect to industrial activities, that pollution encompasses more types of impairment than pollutant discharges alone. Similarly, in section 304(a) of the statute, which requires EPA to develop criteria and information to guide states in the development of water quality standards under section 303(c) of the Act, Congress plainly distinguished between the development of water quality criteria addressing the impact of "pollutants," and "information" on "the factors necessary to restore and maintain the chemical, physical, and biological integrity of all navigable waters." Likewise, in section 510, Congress provided that nothing in the Act precludes or denies other rights of states with respect to "(A) any standard or limitation respecting discharges of pollutants, or (B) any requirement respecting control or abatement of pollution." Obviously in this provision, Congress acknowledged the difference between programs to limit the discharge of pollutants and broader efforts to control other forms of pollution.

Other operative provisions of the statute, however, appear to confuse the issue hopelessly. In section 208, and in a related provision directing EPA to provide information needed to implement section 208, Congress identified as problems both the runoff of pollutants from agriculture, silviculture, mining, construction sources, and so forth as well as other types of pollution such as salt water intrusion and "changes in the movement, flow, or circulation of any navigable waters" due to hydrological modifications. However, in section 319, which is a nonpoint source pollution control initiative added in 1987 and designed, at least in theory, to redress the inadequacy of nonpoint source control programs in the previous fifteen years, Congress again appears to have conflated the distinct ideas of pollution and pollutants. Some aspects of section 319 parallel Congress's recognition in section 101(a)(7) (which was also added in 1987 to establish a goal to control nonpoint source pollution), that nonpoint sources contribute broadly to the pollution of the nation's waters; that is, they alter chemical, physical, and biological integrity in ways that go beyond the runoff of pollutants. For example, section 319(a)(1)(A) directs states to develop reports identifying waters that "without additional action to control nonpoint sources of pollution, cannot reasonably be expected to attain or maintain applicable water quality standards or the goals or requirements of this' chapter."

Other subsections of section 319, however, seem to belie the distinction. Section 319(a)(1)(B), for example, requires states to identify categories of nonpoint sources or particular sources which "add significant pollution [to waters] in amounts which contribute to such portion not meeting such water quality standards or such goals and requirements." Similarly, section 319(b)(1) requires states to develop management programs "for controlling pollution added from nonpoint sources to navigable waters." If the definitions and distinctions between pollutants and pollution are to be taken seriously, it makes no sense to discuss "additions of pollution" and "adding significant amounts" of pollution.

Depending on how it is interpreted, section 303(d) creates similar confusion. This section is the once largely dormant and now famous Total Maximum Daily Load (TMDL) provision requiring states to allocate allowable control obligations among various sources such that aggregate impacts from multiple sources do not cause ambient water quality standards violations. Section 303(d)(1)(A) requires states to identify waters for which the first round of technology-based effluent limitations are not adequate to implement "any water quality standard," i.e., not just those standards involving chemical pollutants. By contrast, under section 303(d)(1)(C), states must establish and implement, for all of those listed waters, the "total maximum daily load... for... pollutants" as a means of remedying the identified violation. If water quality standards are supposed to "serve the purposes of the Act," and hence to identify all types of conditions that impair chemical, physical, and biological integrity, then the listing provision in section 303(d) logically covers all forms of water body pollution. Moreover, as recently confirmed by the Ninth Circuit, section 303(d) would make no sense if only point sources are included in the calculation of total pollutant loadings to an impaired water body. By limiting the effective control mechanisms to total maximum daily loads of pollutants, Congress included in section 303(d) no direct mechanisms to redress other sources of pollution.

A similar dichotomy between the breadth of impaired water body identification listing and the tools designed to address that impairment appears in section 304(1), the so-called "toxic hot spots" provision that was also added in 1987. States are required to identify water bodies impaired specifically with respect to toxic pollutants, and also, as expressly found by the Ninth Circuit, waters that fail to meet much more broadly defined water quality standards. Yet the tools identified and required to redress those problems are limited to "individual control strateg[ies] [designed to] produce a reduction in the discharge of toxic pollutants from point sources... through the establishment of effluent limitations... and water quality standards... which reduction is sufficient, in combination with existing controls on point and nonpoint sources of pollution, to achieve the applicable water quality standard." As with section 303(d), the reach of the remedial tools included in section 304(1) does not match the breadth of problems that must be addressed under the listing provision.

Again, apparent anomalies in the statutory text potentially can be reconciled through the statute's general allocation of responsibility for the control of point source and nonpoint source pollution. Under the plain language of sections 303(d) and 304(1), states are supposed to identify all impaired water bodies regardless of the nature and source of pollution: 1) point sources of pollutants, 2) nonpoint sources of pollutants, and 3) other sources of pollution. The TMDL provision in section 303(d) logically requires states to identify the maximum amount of pollutants from all sources consistent with applicable water quality standards for each impaired water segment, with an appropriate margin of safety, and to allocate those allowable pollutant loadings among point and nonpoint sources. While the states are given the first crack at this mechanism, EPA is directed to correct deficiencies in state efforts. Similarly, allowable loadings of pollutants from point sources can be imposed directly through NPDES permits, which give EPA direct implementing authority absent adequate state action. Efforts to control other forms of pollution apparently are left to other portions of the law, resulting in EPA's ambiguous authority to act if a state fails to do so. Likewise, states are directed to identify all sources of water pollution in the section 304(1) listing process. The specified control mechanisms in that provision ("individual control strategies [designed to] produce a reduction in the discharge of toxic pollutants from point sources") however, were intended to operate "in combination with existing controls on point and nonpoint sources of pollution, to achieve the applicable water quality standard."

Thus, in some respects, the statutory text provides evidence that Congress intentionally designed two sets of statutory purposes of disparate breadth.

Ambiguities and inconsistencies in other provisions, however, cast some doubt on that design. Did Congress distinguish between these terms and ideas knowingly, but occasionally slip in the original text or in subsequent amendments; or did Congress simply use the two terms carelessly and unknowingly, but through sheer statistical chance in ways that made sense in some provisions? In National Wildlife Federation v. Gorsuch, which concluded that water quality changes induced by discharges of river water through dams did not constitute a "discharge of a pollutant" for purposes of the discharge prohibition in section 301(a) and the permit requirement of section 402, the D.C. Circuit found that courts should presume Congress distinguished between the terms "pollutant" and "pollution" intentionally. However, to better understand Congress's intent in drawing these statutory distinctions, it is useful to explore the legislative history of the 1972 Act and relevant subsequent amendments.

Legislative History

Senator Edmund Muskie, the principal Senate architect of the 1972 Act, argued that the statute's goals should have more than the usual weight: "These are not merely the pious declarations that Congress so often makes in passing its laws; on the contrary, this is literally a life or death proposition for the Nation." This rhetoric, of course, can be viewed as little more than a hortatory statement of the statute's hortatory objectives. Similarly, some statements in the legislative history suggest Congress understood its goal as restoring "natural" ecosystem conditions, but with language too general to be of much help.

The 1972 Senate Report, for example, articulated "a clearly set goal of natural water quality" and a "restored and protected natural environment." More specific indications are needed to determine what Congress understood or meant by "chemical, physical, and biological integrity," and by distinguishing "pollution" from "pollutants." There is, however, significant legislative history indicating Congress gave the matter of aquatic ecosystem "integrity" more specific attention. The 1972 Senate Report, for example, evidences an intent to restore aquatic ecosystems to as close as possible to their natural state:

Maintenance of such integrity requires that any changes in the environment resulting in a physical, chemical or biological change in a pristine water body be of a temporary nature, such that by natural processes, within a few hours, days or weeks, the aquatic ecosystem will return to a state functionally identical to the original. In those water bodies which are not pristine, it should be the national policy to take those steps which will result in change towards that pristine state in which the physical, chemical and biological integrity of the water body can be said to exist... an objective which minimizes the burden to man in maintaining a healthy environment, and which will provide for a stable biosphere that is essential to the well-being of human society. This relatively "purist" Senate position is confirmed by the legislative history of later amendments.

The 1977 Senate Report, for example, explained: "A balanced, indigenous population is that which would naturally occur in those particular waters without the effect of man's activities." The 1985 Senate Report went even further, linking the desired state to the absence of "pollution": [T]he term "balanced indigenous population" is defined as the community that can reasonably be expected to occur in a given area absent pollution.

Its existence is generally determined by comparing the ecological community in question to that of nearby healthy communities existing under comparable but unpolluted physical and environmental conditions. The House Committee articulated a more functional view of the term "integrity," based on the concept of preserving natural ecosystem structure and function as opposed to absolute original ecosystem conditions. Nonetheless, this legislative history clearly indicates that the sponsors gave the matter far more than casual thought. The 1972 House Report explains:

The word "integrity" as used is intended to convey a concept that refers to a condition in which the natural structure and function of ecosystems is maintained. Although man is a "part of nature" and a product of evolution, "natural" is generally defined as that condition in existence before the activities of man invoked perturbations which prevented the system from returning to its original state of equilibrium.... This definition is in no way intended to exclude man as a species from the natural order of things, but in this technological age, and in numerous cases that occurred before industrialization, man has exceeded nature's homeostatic ability to respond to change. Any change induced by man which overtaxes the ability of nature to restore conditions to "natural" or "original" is an unacceptable perturbation.

While one could conclude that the House interpretation of aquatic ecosystem integrity is less absolute and hence less "environmental" than that of the Senate, arguably the House Committee's explanation reflects a more sophisticated understanding of modern ecological science than did the Senate's report. If so, perhaps it reflects even more conscious deliberation and intent that those words be taken seriously. These statements from the 1972 House and Senate reports, while reflecting less than perfect agreement, nevertheless indicate that the meaning of the overarching statutory objective was the subject of considered congressional deliberation. Unfortunately, language in the 1972 Conference Committee Report reflects either an even more absolute interpretation than reflected in the Senate Committee report, or serious confusion about the difference between pollution and pollutants.

In explaining section 102 of the Act, the report asserts: "The term 'abating or reducing pollution' has been revised to read 'preventing, reducing, or eliminating pollution.'" Similar explanatory language was offered with respect to section 103: "The term 'abatement' of pollution has been stricken and replaced by the term 'reduction and elimination' of pollution." This change from "abatement" to "elimination" could suggest the Conference Committee was confused and really intended to apply the concept of elimination to the "zero discharge goal" regarding the discharge of pollutants. If the Conference Committee did intentionally mean elimination of pollution as more broadly defined in the text, it would imply the elimination of every type of human alteration of the chemical, physical, and biological integrity of the nation's waters.

Language in the 1972 Senate Report reflects either similar confusion between the zero discharge goal established for discharges of pollutants and the broader but necessarily less absolute task of restoring natural aquatic ecosystem integrity, or a misunderstanding of the breadth of causes of water pollution and the steps needed to address them: The "natural... integrity" of the waters may be determined partially by consultation of historical records on species composition, partially from ecological studies of the area or comparable habitats; partially from modelling studies which make estimations of the balanced natural ecosystem based on the information available. It is expected that the ultimate mechanism for the restoration and maintenance of the natural integrity of the waters will be the complete cessation of discharge of pollutants into waters.

Tentative Conclusions: This evaluation of the text and legislative history of the 1972 Act and relevant subsequent amendments suggests several conclusions regarding congressional intent. First, in the 1972 statute, Congress quite knowingly established a broader objective of "restoring and maintaining the chemical, physical, and biological integrity of the Nation's waters," along with narrower, interim goals of water quality suitable for protecting and propagating fish and wildlife and public recreation. Other statutory goals, such as reducing or eliminating the discharge of pollutants and eliminating the discharge of toxic pollutants in toxic amounts, were designed to further those goals and objectives.

Relatedly, both houses of Congress seem to have given the concept of aquatic ecosystem integrity due deliberation, indicating that the Act's opening phrase was not intended as a mere rhetorical flourish. It does appear the Act's chief sponsors in the House of Representatives and the Senate disagreed on the precise meaning of the statutory objective, as is true of many other provisions of the Act. The Senate Committee's understanding of the concept of integrity seems to suggest a return to pristine, natural ecological conditions, while the House Committee suggested that the concept of integrity implies a return to natural ecosystem structure and functions. Nevertheless, the very existence of that subtle debate suggests the sponsors on both sides of the Capitol took the overriding statutory objective quite seriously.

This apparent intent to take seriously the ecosystem integrity objective is matched by the statute's operative provisions. In particular, Congress clearly had some reason for enacting distinctly different definitions of a "pollutant" as a discrete substance discharged into water, and of "pollution" as all human activities that alter the chemical, physical, and biological integrity of water. Moreover, Congress's use of the terms "pollutant" and "pollution" in the 1972 text seems reasonably consistent (with some exceptions), again indicating the distinction was a considered one. In later amendments, Congress became somewhat less precise in its use of the terms "pollutant" and "pollution," perhaps reflecting that later members of Congress had lost focus on the original distinction between the terms. Congress's shift, however, may have been driven by the fact that EPA and state implementation of the Act had focused almost entirely on the discharge of pollutants, and very little on broader sources of water pollution. This overwhelming focus on implementing the pollutant control aspects of the CWA at the expense of the statute's broader pollution control aspirations suggests it would be useful to examine the degree to which those distinct objectives have been achieved three decades after the 1972 law was adopted.

Achieving the Clean Water act Objective and Goals

In evaluating the nation's progress under the CWA approximately a decade ago, two colleagues and I concluded that significant progress had been made in reducing discharges of pollutants from point sources, but that far less had been accomplished to reduce all forms of pollution from all sources. Recently, as part of a broader effort to evaluate U.S. progress in meeting the sustainable development goal a decade after the 1992 Earth Summit, I revisited those conclusions, and found—unfortunately—that little had changed. The following summarizes briefly the results of that analysis.

Progress in Controlling Pollutants

Available empirical evidence shows significant progress in reducing discharges of pollutants to U.S. waters over the past thirty years, although major problems certainly remain. While there are many legitimate ways to assess effort and progress in this area, several indicators demonstrate the main point. As of 1972, despite decades of significant federal, state, and local investment in public sewage treatment infrastructure, domestic sewage remained a major unaddressed source of water pollution. Tremendous efforts were made and billions of public dollars were invested to address those problems under the 1972 CWA and its later amendments, with dramatic evidence of improvement.

As of 1996, approximately seventy-three percent of the U.S. population was served by public sewage treatment plants, with ninety-nine percent of plants providing secondary treatment or better. Lest we be lulled into complacency, however, this aspect of water pollution control is far from over. According to EPA's most recent analysis, the nation faces estimated public sewerage needs of almost $140 billion by 2016, for new plants, collection systems, and efforts to control sewer system overflows. Similarly, from 1972 through the early 1990s, more than $57 billion was invested in industrial water pollution control, resulting in reductions of millions of tons of conventional pollutants and more than a billion pounds of priority toxic pollutants per year. As with municipal sewage, however, significant releases of industrial pollutants into U.S. waterways continue. For example, U.S. industries reported releases of almost 261 million pounds of toxics into surface waters in 2000, up from 200 million pounds in 1990.

Trends in Ambient Water Quality

Despite significant progress in reducing the discharge of pollutants from municipal and industrial point sources, trends in ambient water quality, i.e., in the "chemical integrity" of U.S. waters, remain ambiguous and inconsistent. This is due in part to the ongoing releases of point source pollutants, runoff of pollutants from nonpoint sources, and other forms of pollution. Large numbers of urban watersheds remain chemically impaired because of pathogens, phosphorus, insecticides, herbicides, and toxics from municipal and industrial sources; and many rural watersheds remain chemically impaired as a result of nutrients, sediment, and agricultural chemicals.

Long-term water quality trends can be difficult to assess because of data gaps, differences in methods of data collection and analysis, and other factors, although methods and information have improved in recent years. Available evidence, however, indicates the nationwide reduction in releases of pollutants from point sources has not universally translated into improved ambient water quality. When consistent, long-term water quality data are available, more often than not they show no clear trends in ambient water quality, with improvements largely in areas associated with point sources, and declining or steady trends in areas associated with nonpoint sources. EPA's Index of Watershed Indicators, a compilation of information on the health of U.S. aquatic resources by watershed, confirms these inconsistent trends. For watersheds for which adequate information exists, EPA reports "more serious water quality problems" in 534 watersheds, "less serious water quality problems" in 794 watersheds, and "better water quality" in 339 watersheds.

Evidence of ongoing impairment of chemical water quality in some areas translates to documented harm to human health and aquatic ecosystems. More than a third of the nation's rivers and almost half of its lakes do not meet designated water quality uses in whole or in part—trends that have not changed significantly over the past decade because of a combination of chemical pollution and other factors. Moreover, disturbing but incomplete data indicate severe ongoing impairment of uses that implicate risks to human health, fish, and wildlife. For example, more than 2500 fish and wildlife consumption advisories were issued in 1998, and more than 2800 were issued in 2000, because of toxics and other contaminants in waterways (mercury, PCBs, chlordane, dioxins, DDT residues), reflecting a steady increase since 1993. (The increasing trend probably is due in part to improved monitoring and reporting.) Similarly, at least 13,410 beach closings or advisories occurred in 2001, some due to heavy rainfall but most due to pathogens and other contaminants—a nineteen percent jump over the previous year and more than twice as many as in 1999 (again, in part due to improved testing and reporting). Because many of these real world indicators relate to the presence of chemical pollutants and pathogens in aquatic ecosystems, they corroborate discharge data that indicate efforts to reduce discharges of traditional pollutants remain incomplete.

Trends in Physical and Biological Integrity

There is considerable and more consistent evidence that the "physical and biological integrity" of the nation's waters has been steadily and seriously declining, with few, if any, signs of better aquatic ecosystem health overall. While we hear a lot about wolves and spotted owls, and without suggesting that those biodiversity issues are not important, aquatic species are declining at a far more alarming rate than are terrestrial species. In a comprehensive analysis of biodiversity in the United States, the Nature Conservancy found that "animals that depend on freshwater habitats—mussels, crayfish, fishes, and amphibians—are in the worst condition overall," and those freshwater-dependent animals have the highest percentages of species characterized as "critically imperiled," "imperiled," or "vulnerable." Breaking down the Nature Conservancy data by watershed, EPA found one aquatic species at risk in 403 watersheds, two to five species at risk in 745 watersheds, and more than five species at risk in 422 watersheds around the country in 1996. The World Wildlife Fund-U.S. (WWF-U.S.) drew similar conclusions in its comprehensive analysis. Although extinctions of individual species are of concern in many ecoregions, WWF-U.S. concluded sufficient numbers of species are at risk that entire "faunal assemblages are in a precarious state."

The fact that aquatic ecosystem health is declining so dramatically on a nationwide basis while evidence of chemical impairment shows inconsistent and ambiguous trends suggests strongly that other major sources of stress are at play. Most aquatic biologists agree that while chemical impairment contributes to the dramatic decline in aquatic ecosystem health, physical and biological impairment is, overall, a more significant factor. In particular, most experts agree that the largest single cause of aquatic species decline is the massive destruction and alteration of all forms of aquatic habitats on a virtually nationwide scale.

Evidence of such habitat impairment is dramatic. For example, Dr. Arthur Benke of the University of Alabama found that all but a single river segment longer that 1000 kilometres (the Yellowstone River in Montana) was substantially altered in some way; of more than 100 river segments longer than 200 kilometres, only 42 remain free-flowing. Similarly, a 1982 analysis by EPA and the U.S. Fish and Wildlife Service (FWS) found that 81 percent of the nation's waters, including more than half of all perennial waters, had fish populations adversely affected by various structural factors.

Other agencies have identified similar levels of impairment of other aquatic habitats over a wider period of time. For example, the Federal Emergency Management Agency reported that floodplain development has destroyed roughly half of the nation's woody riparian habitat. In a 1991 RePort to Congress, FWS found that more than haft of the original wetlands in the coterminous states had been lost through draining, dredging, levying, and flooding, with continuing losses of 260,000 acres per year. The United States Army Corps of Engineers found that more than a half-million stream miles are adversely affected by erosion. Many other stream miles have been inundated, dammed, channelized, dewatered, rip-rapped, and otherwise altered in ways that impair or destroy important habitat. More than 600,000 stream miles have been inundated by literally thousands of dams. Diversions from those dams as well as irrigation canals and other structures have seriously altered natural stream flows and habitats, in some cases leaving little if any water available for fish and other aquatic species.

While some of those evaluations are now decades old, physical impairment usually remains, absent affirmative restoration efforts. Chemical impairment, by contrast, in some cases will dissipate after the source of impairment ceases or subsides. Moreover, recent information shows continued loss and impairment of aquatic habitats, although some restoration and protection programs have succeeded in slowing the rate of those losses. For example, FWS estimates that between 1986 and 1997, an additional 644,000 acres of U.S. wetlands have been lost. The good news, however, is that the rate of wetland loss has slowed from 458,000 acres per year between the mid-1950s and mid-1970s to 290,000 acres per year from the mid-1970s to mid-1980s, and to 58,500 acres per year from 1986 to 1997.

Similarly, according to the recent conservation assessment of freshwater ecosystems conducted by the WWF-U.S., trends in aquatic ecosystem health continue to decline dramatically. WWF-U.S. found that overall, "few [U.S. freshwater aquatic] ecoregions remain intact. A relatively small number are currently so degraded as to be potentially beyond restoration." According to the classifications used in the study, of the seventy-six major aquatic ecoregions in North America only a handful were characterized as "relatively intact" or "relatively stable"; the vast majority of those are in Alaska and northern Canada, places with relatively few anthropomorphic land use changes. The rest are assessed as "vulnerable," "endangered," or "critical."

Likewise, a recent assessment of riparian area health published by a committee of the National Research Council reported widespread hydrologic, geomorphic, and other impairment of riparian habitats, along with accompanying impacts to water quality and aquatic ecosystem health. The report documents as much as ninety-five percent loss of natural vegetation in some riparian areas, "indicating that riparian areas are some of the most severely altered landscapes in the country." As a result, the report concludes the "functions of these riparian areas are greatly diminished compared to what occurred historically."

Aquatic Ecosystem Integrity—an Historical Postscript

The statistics reported above indicate rather irrefutably that the overriding objective of the CWA—to "restore and maintain the chemical, physical, and biological integrity of the Nation's waters"—has not been met. Indeed, we appear to be moving swiftly and disturbingly in the opposite direction. However, a more pointed perspective on the issue can be provided by looking at several "then-and-now snapshots" of aquatic ecosystem health. In the pre-dam era, the Columbia River boasted a salmon population approximately of 10-16 million; today, only about 2 million remain, most of which are hatchery fish. As a result, many stocks of Northwest salmon have been listed as threatened or endangered under the Endangered Species Act. The decline is blamed largely on a combination of habitat loss, timber harvesting, alteration of migration and spawning areas inducing temperature changes and other effects because of the region's massive combination of dams and hydroelectric facilities, and genetic modifications to fish populations as a result of the region's equally massive system of fish hatcheries.

While Pacific Northwest salmon woes have attracted significant attention—perhaps because salmon are tasty, attractive, and symbolic —other historical fish populations have fared even worse. At the turn of the twentieth century, for example, the Illinois River supported a commercial carp fishery of about 15 million pounds per year; today the carp are virtually gone. When Europeans arrived in North America, the Susquehanna River supported a shad fishery that rivaled the Columbia River salmon. Reportedly it appeared as if one could walk across the river on their backs. Shad were caught by the ton in the eighteenth century in nets that blocked entire rivers. But (according to the Maryland State Fish Commissioner) harvests of more than fifteen million pounds per year dropped to a "worrisome" four million pounds in 1876 to less than two million pounds by 1922, and to less than 55,000 pounds in 1978 before the harvest were banned.

Based on these historical anecdotes and comprehensive information documenting serious and ongoing declines in U.S. aquatic ecosystems, the chemical, physical, and biological integrity of the nation's waters has neither been restored nor maintained. We turn next, therefore, to an evaluation of possible explanations for the dramatic gap between the overarching stated objective of the CWA and the stark reality of continuously declining aquatic ecosystem health.

The Unwritten Books in the Water Quality Trilogy

While explanations alone will not solve the problem and reverse the dramatic negative trends in aquatic ecosystem health, it is essential to identify the most likely causes of the ongoing problem in order to design an effective remedial strategy. Possible explanations identified below (which overlap to some degree) include 1) Congress's policy decisions regarding allocation of responsibility for various aspects of water pollution control (federalism), 2) differences in the nature of point and nonpoint source pollution and their most likely control strategies, 3) lack of adequate knowledge and understanding of mechanisms to redress water pollution, as more broadly defined in the CWA, compared to the relatively more straightforward task of treating pollutants in municipal and industrial waste streams, 4) the major gap between the aspiration announced in the opening sentence of section 101(a) of the CWA and effective implementing tools in the rest of the law, and 5) inadequate implementation of the authority that does exist.

Federalism, Water Pollution, and Nonpoint Sources

There are several strong indications in the text of the CWA, some direct and some implicit, that while Congress designed a strong federal role in programs to reduce or eliminate the discharge of pollutants, including both oversight and direct action, it intended that states would be primarily responsible for the more broadly defined problem of pollution. At the outset, two statements of policy in section 101 of the statute, one included in the 1972 law and a second added in 1977, suggest strong deference to states with respect to nonpoint sources and other sources of aquatic ecosystem "alteration." Section 101(b) of the Act proclaims:

It is the policy of the Congress to recognize, preserve, and protect the primary responsibilities and rights of States to prevent, reduce, and eliminate pollution, to plan the development and use (including restoration, preservation, and enhancement) of land and water resources.... It is further the policy of the Congress to support and aid research relating to the prevention, reduction, and elimination of pollution, and to provide... financial aid... [for] the prevention, reduction, and elimination of pollution.

Section 101(g) of the Act, added in 1977, adds: It is the policy of Congress that the authority of each State to allocate quantities of water within its jurisdiction shall not be superseded, abrogated or otherwise impaired by this chapter. It is the further policy of Congress that nothing in this chapter shall be construed to supersede or abrogate rights to quantities of water which have been established by any State. Federal agencies shall cooperate with State and local agencies to develop comprehensive solutions to prevent, reduce and eliminate pollution in concert with programs for managing water resources.

The broader statutory definition of pollution obviously implicates land use, water use, planning, and regulation to a far greater degree than does regulation of discrete discharges of pollutants from point sources. However, the Supreme Court has noted that while the "states' rights" provisions quoted above are intended to preserve state authority with respect to allocation of water rights and other matters, "they do not limit the scope of water pollution controls." Moreover, other provisions in the law shed some doubt on this apparently clean division of authority. Section 102(a), for example, instructs EPA, along with other state and federal agencies, to "prepare or develop comprehensive programs for preventing, reducing, or eliminating the pollution of the navigable waters and ground waters." Similarly, in section 104(a), Congress provides that EPA shall "establish national programs for the prevention, reduction, and elimination of pollution."

Assuming, as we must, that Congress intended "pollution" to have the same meaning in sections 101(b) and 101(g) as it does in sections 102(a) and 104(a), it would appear the states do not have exclusive jurisdiction over the broader issue of "pollution" control.

Neither section 102(a) nor section 104 of the Act contains specific regulatory authority or other tools with which to address the full range of pollution sources directly. Section 102 in particular includes no specific restrictions on what EPA might, in the exercise of its broad, general authority to effectuate the purposes of the statute, include in such "comprehensive" water pollution control programs. Apparently, EPA simply has not chosen to interpret this open-ended but vague provision expansively. Section 104, in turn, contains very specific areas of EPA authority, but those are limited to such non-regulatory activities as research, training, pollution control demonstration projects, technical assistance, expert committees, information dissemination, monitoring, and surveillance. Again, EPA has not chosen to implement this provision broadly.

EPA's reticence to adopt broader pollution control programs under sections 102 and 104 undoubtedly is explained by the contrast between these somewhat vague provisions and more precise parts of the CWA in which EPA's authority is not only specific but mandatory. EPA has direct and exclusive authority in certain aspects of point source pollutant control, such as the establishment of categorical effluent limitations for industrial point sources and secondary treatment, and analogous requirements to control point sources of municipal storm water. In other areas of point source control, such as the issuance of permits to govern otherwise prohibited discharges of pollutants, EPA and the Army Corps of Engineers have direct regulatory authority absent delegation of that responsibility to qualified states with approved programs. Likewise, section 309 of the Act confers on EPA a broad range of enforcement powers relating to violations of point source control obligations, even with respect to requirements imposed by states.

By contrast, programs directed specifically at nonpoint source pollution control—which as argued above encompass a broader range of impairment types as well as different sources of pollutants than those addressed in the point source control provisions of the Act—are ceded more clearly to the states. In section 208 of the Act (included in the 1972 statute) and in the more specific section 319 added in 1987, EPA is given supervisory and program approval authority; but unlike the permitting and enforcement provisions for point sources, EPA lacks direct implementation or regulatory authority in the face of nonexistent or inadequate state implementation. At most, under the nonpoint source control provisions, EPA is authorized to withhold grant funding for delinquent states. This policy judgment appears consistent with Congress's reluctance, as expressed in sections 101(b) and (g) of the Act, to allow extensive federal intrusion into areas of regulation that might implicate land and water uses in individual states. The limited sanction of grant withdrawal, of course, would most likely result in less pollution control rather than more.

The water quality standards (WQS) provisions of the Act occupy a curious middle ground in this spectrum of relative federal and state authority. Consequently, they represent a potential untapped source of federal authority to stimulate better pollution control efforts. As with the point source permitting and enforcement programs, EPA is given clear authority—indeed a nondiscretionary mandate to step in and promulgate legally enforceable water quality standards where states fail to do so, or to do so adequately.

This meshes with the allocation of authority to the extent that the point source control requirements of the Act are designed in part to implement ambient WQS. Implementation of WQS, however, applies to nonpoint as well as to point sources, especially given the reality that the majority of remaining violations of WQS nationally are caused largely or entirely by runoff rather than point source pollution. Violation of a WQS should trigger the listing of impaired waters under the TMDL provision in section 303(d), whether or not that water body is impaired by point or nonpoint sources, and under the nonpoint source assessment provision in section 319(a), which triggers state obligations to develop nonpoint source management programs for those waters under section 319(b). Thus, EPA's authority to adopt enforceable WQS clearly can trigger nonpoint source control obligations, even if EPA lacks the authority to implement those obligations itself. This does not contradict the congressional policy expressed in sections 101(b) or (g), however broadly or narrowly those limitations may be construed. In establishing water quality standards where states fail to do so adequately, EPA's role is merely to set the standards by which impaired waters are identified, and the definition of water quality necessary to meet the statutory objective and goals.

It is not to make basic policy decisions about land and water use that may be necessary to adopt and implement a nonpoint source pollution control program. Likewise, while EPA may have authority to adopt and implement TMDLs under section 303(d) for those waters identified as impaired in whole or in part due to nonpoint sources, and while EPA clearly has the authority to implement those TMDLs with respect to point sources via the point source permitting program, EPA's authority to do so with respect to nonpoint sources is unclear, and the subject of heated dispute. Of course, the federalism policy of the CWA alone does not explain why large numbers of aquatic ecosystems remain physically and biologically impaired, and appear to be declining further, three decades after the 1972 CWA.

It is also necessary to conclude that Congress's experiment of deferring largely to the states to address broader issues of pollution has failed because the states have failed to get the job done absent a stronger federal presence. Indeed, the very contrast between the degree of progress made in controlling discharges from point sources over the past thirty years, compared to the relative lack of success in controlling other forms of pollution, is evidence of the potential merits of a stronger federal-state partnership in the area of water pollution control more generally. If so, it is time for Congress to revisit the allocation of power between EPA and the states with respect to nonpoint source pollution control programs specifically and comprehensive pollution control efforts more generally.

Knowledge, Uncertainty, and Pollution Control

There is little doubt that Congress understood reasonably well in 1972 that the nation's waters were impaired by activities other than traditional municipal and industrial point source discharges. The 1972 Senate Report acknowledged, at least with respect to runoff pollutants from nonpoint sources:

One of the most significant aspects of this year's hearings on the pending legislation was the information presented on the degree to which nonpoint sources contribute to water pollution. Agricultural runoff, animal wastes, soil erosion, fertilizers, pesticides and other farm chemicals that are a part of runoff, construction runoff and siltation from mines and acid mine drainage are major contributors to the Nation's water pollution problem. Little has been done to control this major source of pollution.... It has become clearly established that the waters of the Nation cannot be restored and their quality maintained unless the very complex and difficult problem of nonpoint sources is addressed.

It is not clear, however, whether Congress specifically addressed, or even considered, the many other ways in which aquatic ecosystem integrity is altered by human activity, despite the fact that such activities fall within the plain meaning of the statutory objective and the related definition of pollution. Nor is it clear whether Congress considered very carefully how those forms of pollution might be addressed by state or federal control programs. What is relatively clear, however, is that in 1972 Congress understood quite well how much was not understood about some of the broader challenges in pollution control. The 1972 Senate Report admonished: "In order to restore the natural chemical, physical, and biological integrity of the nation's waters, a great deal more must be learned about the nature and functioning of aquatic ecosystems and the impact of improper land use development." Likewise, the Report noted:

> *In view of the objective of the Act to restore the natural integrity of the nation's waters, the Committee has included a new subsection 104(m) authorizing the designation of River Study Centres... [that] could apply existing and developing knowledge to the problem... of growth and development within river basins and the impact of that development on the best use of water resources....*

By that time, the science of water pollution control engineering and technology as applied to municipal and industrial point sources was well established, although significant improvements were still to come. Far less was known, however, about methods to control nonpoint source runoff and other forms of pollution as defined in the new law. To address these gaps in information and understanding, the 1972 Act and later amendments included numerous provisions aimed simply at basic and applied research and development, demonstration projects, and education and training. It is entirely possible that Congress elected not to adopt more prescriptive provisions to control sources of water pollution other than discharges of pollutants because it lacked information about the nature of various sources of pollution, their impacts on aquatic ecosystems, how such programs might be structured, and what restoration and control methods would be used. If that is the case, however, given another three decades of research under myriad CWA and other programs, and given our substantially increased information at least about the significant harm that continues to occur because of those pollution sources, it is time for Congress to revisit that judgment.

Aspirations and Reality

As noted at the outset, in passing environmental statutes Congress sometimes articulates lofty aspirations but fails to provide EPA or other agencies with adequate authority to accomplish those goals. Alternatively, in some statutes Congress provides such a wide range of discretion that the difficult decisions are left to the agencies. In both cases, Congress can proclaim publicly that it has addressed an important environmental issue but has delegated to the agencies the difficult judgments about who must bear the costs or benefits of action or inaction.

In the CWA, Congress suggested to the American public thirty years ago that the chemical, physical, and biological integrity of the nation's waters would be restored and maintained. At a minimum, through the CWA goals Congress intimated that all waters would be protected for healthy fish, aquatic life, and public recreation by no later than 1983. Based on the available empirical information outlined above, however, the statute's interim water quality goal has been met only in part, while for most aquatic ecosystems the overriding ecosystem integrity objective of the law seems to be just a fading dream. One possible explanation for this failure is that Congress simply failed to match its laudable rhetoric with adequate implementation tools and authority.

This theory has significant support in the text of the statute, which suggests a decided pattern in how the distinct issues of pollution and pollutants are addressed. Throughout much of the statute, the broader issue of pollution is addressed in the comprehensive but largely unenforceable, voluntary planning provisions. By contrast, most of the more precise, enforceable provisions of the Act apply to discharges of pollutants.

A particularly interesting example of this distinction appears in section 304(a), under which EPA develops the background information used by states and EPA to adopt enforceable water quality standards. Section 304(a)(1) directs EPA to develop and publish water quality "criteria" regarding "the kind and extent of all identifiable effects on health and welfare... from the presence of pollutants in any body of water," along with the "concentration and dispersal of pollutants, or their byproducts, through biological, physical, and chemical processes," and on the "effects of pollutants on biological community diversity, productivity, and stability." Section 304(a)(2), by contrast, instructs EPA to develop and publish... information (A) on the factors necessary to restore and maintain the chemical, physical, and biological integrity of all navigable waters... (B) on the factors necessary for the protection and propagation of shellfish, fish, and wildlife... and to allow recreational activities in and on the water....

Why was EPA instructed to produce "criteria" for the effects of pollutants but "information" regarding the "factors necessary" to attain the ultimate aquatic ecosystem integrity objective of the CWA and the interim goal of protecting aquatic life and public recreation? One possible explanation, consistent with the inadequate information theory presented in Part IV.B, supra, is that Congress assumed pollutant discharges were amenable to quantifiable criteria, while the broader objective of chemical, physical, and biological integrity could be addressed largely or entirely through more qualitative information. Water quality criteria themselves, however, can be expressed in narrative as well as numeric forms. It is also possible that this distinction reflects the pattern, in which Congress intended point source discharges of pollutants to be regulated via relatively precise, enforceable technology-based and water quality-based effluent limitations, and that the latter would be driven by specific, enforceable water quality criteria adopted by the states or by EPA.

Pollution, by contrast, was to be addressed in planning programs designed to be more comprehensive in the sense that they would address a broader set of issues but with states rather than EPA deciding the key policy decisions regarding who, what, and how to regulate. The appropriate role for EPA in the latter process, perhaps, would be to provide information rather than more specific criteria. There are, however, several exceptions to this apparently sharp line between precise, enforceable provisions to address discharges of pollutants and the generally fuzzier programs to address broader issues of pollution. These exceptions suggest other sources of untapped statutory authority for the states, EPA, or both to address issues of pollution more widely and more effectively. Section 401 of the statute, for example, which authorizes states to impose a broad range of water quality conditions on activities subject to federal licenses and permits, has been interpreted in a manner that could help bridge the gap between regulation of pollutants and pollution. Pursuant to section 401(a)(1), water quality certification authority applies to "[a]ny applicant for a Federal license or permit to conduct any activity... which may result in any discharge into the navigable waters." This authority could be interpreted to apply to activities other than point source discharges because of the statutory distinction between "discharge" and "discharge of a pollutant," although that reading was rejected—in my view incorrectly—by one court of appeals. Perhaps more important, in PUD No. 1 of Jefferson County v. Washington Department Ecology (Jefferson County), the U.S. Supreme Court interpreted section 401(d) to allow states, once a discharge is found, to impose requirements designed to address water pollution, and not just those related to the discharge of pollutants. Indeed, in upholding conditions of a state water quality certification that established minimum instream flow conditions for a proposed hydroelectric facility, the Court specifically invoked the statutory objective of "chemical, physical, biological and radiological integrity," and relied expressly on the broader statutory definition of pollution:

[T]he Act's definition of pollution as the "man-made or man-induced alteration of the chemical, physical, biological, and radiological integrity of water" encompasses the effects of reduced water quantity. 33 U.S.C. [section] 1362. This broad conception of pollution—one which expressly evinces Congress' concern with the physical and biological integrity of water—refutes petitioners' assertion that the Act draws a sharp distinction between the regulation of water "quantity" and water "quality."

Moreover, the Court reasoned that certification requirements were valid if designed to ensure compliance with any valid state water quality standards, including designated uses and antidegradation requirements as well as specific numeric water quality criteria. While the Court indicated that state authority under section 401 "is not unbounded," apparently this provision can be used by states, through enforceable conditions on federal licenses and permits, to address any form of pollution as broadly defined in the Act, and to ensure compliance with any aspect of a valid water quality standard, including those designed to effectuate the broadly stated statutory objective of "chemical, physical, and biological integrity." It is significant, of course, that this specific, enforceable authority to redress pollution in addition to pollutants is given to the states rather than the federal government. Moreover, section 401 allows states to address broader sources of pollution related to federal licenses and permits only where that authority is triggered initially by a discharge of some kind. However, once such a condition is included in a water quality certification, it is independently enforceable in a citizen suit brought under section 505. Another significant provision of the statute confers direct authority to federal agencies to address physical and biological integrity with respect to some types of activities with the potential for severe "alteration" of the aquatic environment. Section 404 authorizes the Secretary of the Army to issue permits "for the discharge of dredged or fill material into the navigable waters at specified disposal sites."

Disposal sites are to be selected in accordance with criteria issued by EPA under section 404(b)(1), based on similar criteria issued for discharges to coastal waters. Moreover, EPA is authorized to prohibit such discharges of dredged or fill material, in consultation with the Secretary, upon finding "that the discharge will have an unacceptable adverse effect on municipal water supplies, shellfish beds and fishery areas (including spawning and breeding areas), wildlife, or recreational areas."

The authority provided in section 404, of course, in one sense continues the general pattern in the rest of the statute because it confers direct federal authority only with respect to discharges into waters of the United States of a specified category of pollutants (dredged and fill materials). However, compared with its sister permitting provision in section 402, which applies to discharges of other types of pollutants, section 404 more expressly directs the agencies to consider potential impacts on aquatic ecosystem health.

Moreover, because discharges of dredged or fill material requiring a section 404 permit often are necessary to conduct other activities that cause pollution of aquatic ecosystems more broadly defined, such as the creation of a reservoir, diversion of water from a river, or channelization or armoring of stream banks, that provision provides direct federal agency jurisdiction and approval or veto authority over those activities as well. The focus of section 404 relative to section 402 probably reflects the realization that such discharges—unlike liquid waste streams from most municipal and industrial point sources—are inherently more likely to affect the physical integrity of aquatic ecosystems on a broad scale. Indeed, such discharges often eliminate portions of the aquatic ecosystem entirely, by converting wetlands and other bodies of water—or portions thereof—into dry land. It is hard to conceive of a more serious "man-made or man-induced alteration of the chemical, physical, biological, or radiological integrity of water" than complete physical elimination; thus, such activities appear to constitute pollution per se.

The extent to which the federal agencies have actually implemented this authority is addressed in the following section. Latent but potentially broader-reaching federal authority over issues involving pollution may reside in EPA's authority to provide water quality criteria guidance under section 304(a) and to adopt water quality standards under section 303(c) of the Act. Because WQS are supposed to "serve the purposes of this chapter," it is certainly permissible if not mandatory that they address the full range of chemical, physical, and biological impairments addressed in the Act's opening objective, as well as the policies articulated in its subsidiary goals. Where a state fails to adopt WQS adequate to address all of those purposes, such as the biological standards for which EPA has adopted significant guidance under section 304(a), EPA has the authority, if not the duty, to step in and do so for the state.

Adoption of WQS, of course, does not directly redress water pollution problems. However, WQS that address the full range of chemical, physical, and biological impairment causing serious aquatic ecosystem declines can be extremely useful in driving other CWA programs. More comprehensive WQS, including biocriteria, can be implemented and enforced directly with respect to point source permits issued under sections 402 and 404 by the states, EPA, and the Army Corps of Engineers, and enforced independently via citizen suits.

They can also be used by states to condition federal licenses and permits for discharges under section 401(a), the full scope of which will turn on the debate about whether that provision encompasses both point and nonpoint sources of pollution. Moreover, under section 313 of the CWA all federal agencies must comply with WQS with respect to "any activity which may result in the discharge or runoff of pollutants." Actions can be brought under section 706 of the Administrative Procedure Act alleging that federal agency actions would cause or contribute to violations of water quality standards.

Water quality standards are also relevant to identifying and listing impaired waters under sections 303(d), 305(b), and 319 of the Act, and of the accompanying programs to redress that impairment. EPA, in turn, must adopt TMDLs for all waters identified under section 303(d) where the state fails to do so, regardless of the nature of the impairment leading to that listing. While the extent to which EPA has the authority to implement those TMDLs remains a subject of heated debate, and while EPA lacks authority to implement nonpoint source pollution management plans under section 319(b), it does have the authority to disapprove state plans under sections 303(e) and 319(b), and to withhold federal grant funding as a result. Thus, EPA's authority to adopt state WQS that address a broader range of chemical, physical, and biological impairment issues has the potential to assert a strong influence on relevant state programs.

In short, while EPA and other federal agencies have some statutory authority to address issues of pollution directly, that authority is limited to particular sources of pollution or to specific steps in the overall statutory program. Accomplishing the statutory objective may have been, as proclaimed by Senator Muskie, a "life or death proposition for the Nation," but Congress provided EPA with few direct lifesaving tools. Instead, much of the rescue effort was left to the states. However, given that EPA does have some statutory authority to act directly, or at least to prod the states more effectively, it is useful to evaluate the extent to which this authority has been used.

Implementation: EPA as Reluctant Master

Despite EPA's mandate in section 102 of the CWA to "prepare or develop comprehensive programs for preventing, reducing, or eliminating the pollution of the navigable waters and ground waters," it is commonly recognized that EPA has devoted the vast majority of its time, effort, money, and regulatory clout to point sources as opposed to nonpoint sources; pollutants as opposed to pollution; chemical as opposed to physical and biological integrity; and technology-based as opposed to water quality-based programs.

While there is evidence in the legislative history that some of the statute's chief sponsors intended that EPA focus initially on implementation of technology-based limitations on point sources, it is hard to imagine that "initially" meant thirty years. In fact, evidence—and perhaps an expression of impatience—that Congress imagined swifter action with respect to other areas of pollution appears in the last articulated statutory goal, added in 1987, that nonpoint source pollution control programs "be developed and implemented in an expeditious manner so as to enable the goals of this chapter to be met through the control of both point and nonpoint sources of pollution."

Perhaps the best indication of EPA's relative lack of focus on physical and biological integrity is that, in hundreds of pages of EPA implementing regulations, concepts of physical and biological integrity, and requirements to control pollution as opposed to discharges of pollutants, are barely present. EPA does reiterate the statutory definition of pollution in its rules, but no regulatory implications appear to attach to that definition. Moreover, EPA affirmatively removes the full water quality trilogy from the operative language of its water quality regulations. EPA rules define "serve the purposes of this Act" as restoring "wherever attainable... water quality for the protection and propagation of fish, shellfish and wildlife and for recreation in and on the water," (the interim "fishable and swimmable" goal) without any reference to restoring and maintaining the "chemical, physical, and biological integrity of the Nation's waters" (the overriding objective of the statute).

Whether or not this omission was intentional, it is significant in terms of the potential scope of the WQS program. If WQS are required to "serve the purposes of this Act," and if those purposes logically encompass the overriding aquatic ecosystem integrity objective of the law, then WQS are required to address all parts of the CWA trilogy. By extracting the statutory objective out of the definition of "serves the purposes of this Act," EPA has weakened the required scope of the program, and all of the statutory and regulatory ramifications that flow therefrom.

EPA's relatively narrow traditional and ongoing CWA focus is probably explained in part by sheer history and inertia, and in part by the magnitude and complexity of the point source problem itself. While Congress may have intended that EPA focus initially on technology-based approaches to controlling point sources programs, the entire structure of EPA's CWA programs were shaped accordingly. It is, after all, often difficult for massive bureaucratic ships to change direction. Moreover, any illusions that the point source problem would be tackled quickly and successfully were grossly optimistic. Those efforts turned out to be vastly more complex, time-consuming, and contentious than the sponsors of the 1972 Act may have believed.

EPA's relative inattention to the broader issue of pollution is also explained in part by the agency's actual lack of statutory authority to address pollution as opposed to discharges of pollutants, but may also be explained in part by EPA's exaggerated perception of its lack of statutory authority, fueled by obvious political barriers to taking on the even more difficult and even more contentious issues of pollution. This explanation is supported by the fact that even where it has some lifesaving tools, by and large EPA has been afraid to jump into the water.

For example, EPA could expand the focus of the point source program itself. Federal courts have indicated repeatedly that, to fulfil the remedial purposes of the statute, the statutory definition of point source should be interpreted broadly. When EPA originally attempted to exclude entire categories of point sources from the NPDES program, environmental groups successfully challenged that practice. The D.C. Circuit Court of Appeals suggested, however, that EPA has limited authority to determine the proper scope of the NPDES program through its regulatory interpretation of the terms "point source" and "nonpoint source." EPA has used that technique to "define" away a large number of agricultural, silvicultural, aquacultural, and other activities that otherwise would constitute point sources under the plain language of the CWA. EPA could pull more polluting activities within the more rigorous control of the point source program by revisiting those definitions.

Similarly, while National Wild life Federation v. Gorsuch upheld EPA's decision not to regulate discharges from dams under the NPDES program, the D.C. Circuit indicated quite clearly that it did so only in deference to EPA's interpretation of an ambiguous statute and legislative history. In fact, the court signalled that had EPA reached the opposite conclusion, that result likely would have been upheld as well. EPA's direct authority under NPDES, of course, is to regulate the discharge of pollutants. However, just as the activities regulated under section 404 also happen to cause significant pollution more broadly defined, the types of pollution sources that might be embraced by a broader EPA definition of point sources also contribute heavily to the types of physical and biological impairment that have escaped significant attention under the CWA.

Federal agencies also have been reluctant to take full advantage of their relatively clear authority to protect aquatic ecosystem integrity under section 404 of the CWA. In its section 404(b)(1) guidelines, EPA did promulgate rather extensive regulations defining the circumstances under which proposed discharge sites should be accepted or rejected. Moreover, those regulations deal extensively with the types of physical and biological integrity that are given far less attention in implementation of other parts of the statute, and could perhaps serve as a model starting point for more extensive attention to those issues in other programs. However, there is serious doubt as to the effectiveness with which those requirements have been implemented in the section 404 program. While the Corps of Engineers has certainly denied fill applications in some cases based on the EPA regulations, the vast majority of requested fill activities under section 404 have been authorized notwithstanding the fact that the discharge of dredged or fill material not only impairs chemical integrity but most often destroys entirely the physical and biological integrity of the affected waters. As a result, while the acreage of wetlands filled has slowed in recent years, it still continues at a disturbing rate. EPA, in turn, has exercised its authority under section 404(c) to veto section 404 discharge sites in some cases. However, that authority has been exercised on only a handful of occasions in the thirty-year history of the CWA.

Even in those areas in which the federal role is primarily one of oversight and approval rather than direct authority to implement, EPA has been relatively timid. Despite three decades of state implementation of nonpoint source pollution control programs under sections 208 and 319, through which Congress presumably intended the states to address broader issues of pollution and not just runoff of chemical pollutants, progress has been slow to nonexistent. Evidence from thirty years of water quality monitoring indicates that water quality in areas impaired largely by nonpoint sources has not improved measurably.

The fact that the physical and biological integrity of the nation's waters continues to decline severely, as measured by indicators of habitat loss and degradation as well as by direct effects on aquatic and water-dependent wildlife species, indicates that under existing state programs even less progress is being made to address issues such as hydromodification, instream flow depletion, channelization, and other activities that severely alter the physical and biological integrity of aquatic ecosystems.

EPA has a clear statutory duty to disapprove state programs that fail to address those problems, based on continued violations of water quality standards and other requirements of the Act. Most notably, if EPA determines that a state nonpoint source pollution management program under section 319 "is not likely to satisfy, in whole or in part, the goals and requirements of this Act," or if "the practices and measures proposed in such program... are not adequate to reduce the level of pollution in navigable waters," EPA must notify the state of necessary program improvements, and approve or disapprove those changes.

Despite the general failure of state programs to meet these basic requirements on a national scale, all fifty states are proceeding happily under EPA-approved programs. While the solitary formal penalty for disapproval may be the withdrawal of relatively small program grants, EPA has not even resorted to this remedy. At a minimum, such action by EPA would highlight more clearly to Congress and the public that the current, decades-old, largely state-driven process for addressing water pollution in the United States is not working. Moreover, the strategic use of EPA's disapproval authority, beginning with the most obviously deficient programs, might stimulate somewhat better state efforts through sheer stigma.

Thus, even where statutory authority exists, in many areas of CWA implementation EPA has given the issues of physical and biological integrity little serious attention compared to its extensive efforts to address chemical pollution. However, one encouraging area in which EPA has moved more decidedly in the direction of improving efforts to focus on physical and biological integrity of the nation's waters has been in the development of biological water quality standards and criteria (biocriteria).

Implementation: The Latent Potential of Biological Water Quality Standards

Biological water quality standards and criteria (biocriteria) define the biological health of aquatic ecosystems by comparing the biota in a given system with those in the least-altered natural system of the same or similar type, using data from multiple trophic levels.

More specifically, EPA defines biocriteria as "narrative or numeric expressions that describe the reference biological integrity (structure and function) of aquatic communities inhabiting waters of a given designated aquatic life use. Biocriteria are based on the numbers and kinds of organisms present and are regulatory-based biological measurements."

EPA explains further that biocriteria are "derived from biological assessments involving integrated measures—indices—of the composition, diversity, and functional organization of a reference aquatic community," which ideally reflect "unimpaired or minimally impaired conditions." Notably, this focus on the natural structure and function of aquatic ecosystems directly parallels the manner in which the House Committee explained what it intended by "integrity" in the original 1972 Act. Thus, efforts to develop, adopt, and implement biological standards and criteria reflect the most serious work to date to address directly this major aspect of the overarching goal of the 1972 CWA.

EPA itself underscores the fact—argued implicitly throughout this Chapter—that a focus on chemical and physical water quality parameters alone cannot meet the goals and requirements of the CWA fully:

[B]iological assessments and adoption of biological criteria must become an equal component of water quality management programs along with chemical, physical, and toxicity based water quality standards. By themselves, traditional chemical, physical and toxicity assessments cannot fully answer questions about the ecological integrity of a water body, or determine whether aquatic resources are being protected.

EPA's conclusion results from the fact that other types of criteria, such as numeric criteria for individual pollutants and even whole effluent toxicity criteria, are designed to identify (and prevent) discrete impacts from particular causes, while biological assessment methods measure the overall health of the resident biological community in response to all sources of impairment combined. Those states that have used biocriteria in conjunction with other forms of water quality criteria have detected ecosystem impairment in situations where using traditional measures alone would have indicated full use attainment; thus indicating that many existing state monitoring and assessment efforts—and accompanying listing efforts under sections 303(d), 319(a), and other provisions of the Act—likely understate the nature and magnitude of the problem. Unfortunately, as with other CWA efforts to address aquatic ecosystem integrity, progress in this initiative has been slow, although steady. The pioneering scientific research that underlies the development of biological standards is more than two decades old. Some states, Ohio in particular, began to develop and use biological indicators in their water quality programs in the mid-1980s, and EPA began developing guidance for other states by the late 1980s.

By 1991, when EPA published its first assessment of the degree to which other states were following suit, relatively few had done so. By 1995, when EPA conducted its first comprehensive survey of state bioassessment programs and uses, forty-one states used bioassessments for some purposes, twenty-nine had developed narrative biocriteria, but only one state had adopted numeric criteria.

EPA's most recent and more comprehensive survey, conducted in 2000 and 2001, indicates that fifty-seven out of sixty-five entities (including states, tribes, and interstate agencies) use some form of bioassessments, but still only twenty-nine have now adopted narrative biocriteria, with eleven more "in a development phase," and still only four states have adopted numeric criteria, with another nine "under development."

Absent development and use of such methods, states and EPA cannot even determine fully and adequately the degree to which the biological integrity of waters being maintained and protected, as required by several of the Act's assessment, reporting, and listing provisions, much less devise and implement actions to protect biological integrity. This slow rate of state adoption of biocriteria is explained in part by EPA's strategy of encouraging rather than requiring state action in this area. But EPA has the legal authority, if not the duty, to require state adoption of biological water quality standards as it has done for other forms of water quality standards, to effectuate the CWA properly, and to adopt federal standards for states that fail to do so. As water quality criteria are required to protect all designated uses in the WQS, including aquatic life uses, and more broadly to "serve the purposes of" the Act, which include restoring and protecting aquatic ecosystem integrity. Moreover, the statute provides that EPA "shall promptly [promulgate WQS]... in any case where... a revised or new standard is necessary to meet the requirements of [the Act]." And as noted above, EPA has expressly concluded that other forms of WQS alone are not sufficient to measure and define the aquatic ecosystem integrity objective of the Act. To fulfil the requirements of the statute properly, especially given slow movement in many states, EPA should require rather than encourage the development and adoption of biological water quality standards. Even slower progress is apparent in the use of biocriteria for purposes other than monitoring and assessment. In its initial program guidance, EPA indicated that biocriteria also could be used for program planning and management and for regulatory purposes. Better information on the location and sources of water body impairment can be used to design and implement more effective watershed protection and nonpoint source pollution control programs, and trend information using those tools can be used to assess the effectiveness of those programs over time.

For example, EPA could use improved information from biomonitoring and assessment to determine whether state nonpoint source management programs are serving the purposes of the Act, and to disapprove or require modification of inadequate programs. Moreover, biological water quality standards can be used, and have been used, to justify permit denials; support the development and application of NPDES permit limits; enforce permit requirements through citizen suits or otherwise; challenge federal agency decisions under section 313 and the APA; and support CWA enforcement actions. Indeed, once biocriteria are adopted as enforceable state water quality standards, NPDES permits must ensure that such standards are met just like any other WQS.

While EPA has articulated a goal of using biocriteria for program implementation and enforcement as well as in regulatory programs, where they exist at all biocriteria have been used largely for assessment and listing purposes, and only sporadically for more specific regulatory tasks. Thus, there is significant untapped potential for adopting and using biocriteria in ways that could significantly fill the gap between the aspirations for biological integrity and the sad reality of ongoing and worsening aquatic ecosystem impairment over the past thirty years since Congress first proclaimed the 1972 anti-pollution objective.

Toward Improved Focus on Physical and Biological Integrity

The text and legislative history of the CWA make clear that Congress intended to address not only chemical water quality, but also the ecological integrity of the nation's aquatic ecosystems. At least in principle, Congress meant what it said when it established, as the overarching objective of the Act, the restoration and protection of the chemical, physical, and biological integrity of the nation's waters. The actual history of CWA implementation, however, along with overwhelming evidence that the integrity of our aquatic ecosystems continues to decline dramatically, suggests that attainment of this objective requires a serious change in direction.

In large part, this serious failure to meet the basic ecosystem integrity goal of the CWA may be attributed to Congress's assignment to the states of the primary responsibility for controlling pollution from activities other than point source discharges of pollutants. Stated differently, by and large Congress simply failed to match its ecosystem integrity aspirations with sufficient, hard legal mandates for EPA and other federal agencies to ensure states address the full range of impairments to the ecological integrity of our waters—or, alternatively, to do so through direct federal action. This failure is demonstrated by the wide gap between the success of point source control programs, over which the Act provides significant federal oversight and direct implementing authority when states fail to get the job done, and the relative failure of state-driven nonpoint source pollution control programs.

Moreover, after thirty years, it is naive to expect that this problem will be solved simply by exhorting states to do a better job, whether through the broader state authority suggested by the Supreme Court in Jefferson County, additional grant funding, better technical assistance, or other incentives. Thus, if we are serious about meeting the Act's ecosystem integrity objective, relief must come from one of two sources, or both. First, the above analysis identified a number of ways in which EPA could make better use of its existing authority to control pollution as opposed to just pollutants, and to control some pollutant discharges more directly.

For example, EPA could exercise its authority under sections 403 and 404 more comprehensively to protect a broader range of important habitats from physical and biological impairment (or outright destruction). It could modify its past regulatory interpretations to expand the scope of activities regulated as point sources. EPA could also require states—rather than merely encourage them—to strengthen biocriteria programs and means to translate those criteria into enforceable, regulatory programs. In particular, EPA has legal authority under section 303(c) to mandate, rather than encourage, state adoption of biocriteria, and to adopt federal standards where states fail to do so; or to do so adequately.

In turn, EPA can use improved biocriteria to expand the impaired water body listing programs under sections 303(d), 305(b), and 319(a) of the Act, and to guide programs to remedy those problems more effectively. EPA has the legal authority, if not the mandate, to require that biocriteria be enforced with respect to all point source discharges through NPDES permits; and states could also use biocriteria more effectively and expansively in section 401 water quality certifications. However, even with these regulatory changes it is unlikely that efforts to restore and maintain the physical and biological integrity of the nation's waters will improve comprehensively. For example, statutory authority is sparse for implementing and enforcing biocriteria with respect to activities other than point source discharges. Thus, in addition to changes at the administrative level, it is likely that statutory amendments are needed as well to close the serious gap between aspirations and reality.

A number of potential amendments come readily to mind. For example, Congress could amend the statute to provide more direct regulatory authority to control pollution as opposed to discharges of pollutants. Congress could amend section 319 to look more like other CWA programs, giving EPA the authority to adopt and implement nonpoint source pollution control programs when states fail to do so. Congress could modify section 303(d) to clarify EPA authority to include implementation plans in TMDLs when states fail to do so, or to do so adequately. Similarly, it could amend section 303(d) to clarify that when waters are listed because of pollution as opposed simply to pollutants, TMDLs should include metrics designed to address those problems. The definition of discharge could be amended to include forms of pollution other than discharges of pollutants, and section 401(a) could be expanded to apply to all federal licenses and permits for activities that cause or contribute to pollution.

Other possible statutory amendments could apply with even greater breadth. For example, Congress could amend section 301(a) to provide: "Except as in compliance with this section and sections 302, 303, 304, 306, 307, 318, 402, and 404 of this Act, the pollution of any water of the United States, including by the discharge of any pollutant, by any person, shall be unlawful." Similarly, sections 309 (enforcement) and 505 (citizen suits) could be expanded to allow suits to enjoin any person from conducting any activity that "causes or contributes to pollution of any water of the United States." In 1977, only five years after articulating the objective of the 1972 Clean Water Act, Congress speculated that "it may be that the States will be reluctant to develop [adequate] control measures... and it may be that some time in the future a Federal presence can be justified and afforded."

Now, another twenty-five years later, the time is long overdue for Congress to revisit that prediction. Indeed, based on the trends in aquatic biodiversity and other factors, if serious action is not taken soon to reverse the decline in U.S. aquatic ecosystems, it does not seem unduly hyperbolic to suggest it may well be too late to do so. The appropriate question is not whether a stronger federal government can be "justified and afforded," but whether we can afford not to adopt and implement more serious federal efforts to address all forms of water pollution, and in short, to write all three books in the water quality trilogy that Congress outlined thirty years ago.

A Watershed Issue: the Role of Streamflow Protection in River Basin Management

The recent crash of salmon, steelhead, and trout populations has drawn attention to the altered state of Northwest Rivers. Columbia Basin salmon runs once numbered perhaps sixteen million fish; today, they hang by a thread. The issue of how to save the salmon has become exceedingly large, complex, and contentious, possibly more so than any other issue in the region. These ancient species are extremely important to the economy, ecology, and culture of the Northwest, and their decline raises fundamental questions about the problems confronting the region's rivers and the plans to restore these rivers to health. While hydropower dams may be the biggest single problem on the Columbia and Snake Rivers, dams are only part of the trouble with rivers in the Northwest.

Many fish stocks are in serious peril even without the harmful effects of large dams. These stocks include resident fish such as the bull trout (which allegedly deserves an Endangered Species Act listing) as well as salmon runs on coastal streams. The region must look beyond dams if it is to restore its fish populations and rivers. There is increasingly wide recognition of the need to look comprehensively at the problems facing rivers, from excessive water temperatures to nonpoint source pollution to dewatering. Much recent attention has been directed to the concept of the "watershed"—the entire basin drained by a particular river or stream—and to the need to address the whole range of factors affecting watershed health. Today, people ranging from federal agency heads to family farmers are discussing watershed planning, watershed restoration, and watershed management.

This focus on watersheds fits within the broader context of "ecosystem management," an approach to natural resources management that considers a broad range of ecological as well as societal factors. In other words, the watershed/ecosystem approach promises comprehensive resource management that takes account of both natural and human values—a truly holistic stew. Conceptually, comprehensive-resource management at the watershed or ecosystem level makes a good deal of sense. In the Northwest, however, this concept is confronted with a very serious and difficult problem: state water law and the prior appropriation doctrine.

Water is the key ingredient in healthy, productive, functioning ecosystems. Especially in the arid and semi-arid country east of the Cascades, water is life itself. Water nourishes all manner of organisms that inhabit rivers and riparian corridors. Water is also biologically and economically crucial to human survival. Any attempt to implement truly comprehensive "ecosystem management" must include management of water to meet both human and environmental needs. In the Northwest, however, water is not managed that way. Instead, water is governed by a system of appropriative rights that allows private users to take all the water out of a stream to meet their needs, leaving nothing for the needs of other people or the environment. Thus, state water law does not fit well within a comprehensive approach to managing resources that considers the interests of all people and all species in maintaining sustainable ecosystems. Rather, northwestern states manage water under the simple directive of prior appropriation: "first in time, first in right."

For this reason, many Northwest streams are completely dried up at certain times of year; many others are drastically lowered, with major impacts on water temperature, pollutant concentrations, the quantity and quality of aquatic habitat, and other values. Most of the water diverted and consumed goes to irrigated agriculture. Lacking adequate water, many Northwest rivers cannot support the fish, birds, plants, and other creatures that depend on them. A watershed is simply not healthy if all the water is removed from its rivers. The four major Columbia Basin states (Idaho, Montana, Oregon, and Washington) have taken steps to protect some of the remaining flows in certain rivers. For most reaches of most rivers, however, instream flow protection is nonexistent or inadequate to protect aquatic life.

Moreover, neither streamflow protection measures nor state water planning pro grams are well connected with many "watershed management" activities of government agencies and local groups, although some notable exceptions exist. For both legal and political reasons, watershed management efforts generally concentrate more on land use and riparian improvements than on water quantity. However, such efforts cannot be fully comprehensive or successful unless they effectively address the need for instream flows. State laws provide for instream flow protection and water planning at the river basin level, but these measures often are not connected with watershed management efforts. Streamflow protection, however, can be integrated into the watershed approach and may be crucial to the success of comprehensive watershed management.

Watershed Management

Throughout the Northwest, government agencies at every level, as well as many private citizens, are beginning to advocate a watershed approach to natural resource management. The following section identifies some of the better-known "watershed management" efforts in the region and discusses some of the reasons why the watershed approach has be come popular.

Watershed Management Efforts by Various Agencies

"Watershed management" comes in all shapes and sizes, from multi-agency efforts to manage all federal forest lands west of the Cascades to volunteer watershed councils that concentrate on one small river basin. The following pages sample such activities in the Northwest.

Federal Efforts

U.S. Forest Service/U.S. Bureau of Land Management (BLM) : Federal natural resource agencies in the Northwest have embarked on some ambitious and high-profile attempts at ecosystem management. The key agencies, the U.S. Forest Service and BLM, manage much of the land in the Columbia River Basin. What ecosystem management actually means to these agencies is somewhat difficult to comprehend. Whatever the concept may mean, the federal land management agencies seem committed to it, as the theme of ecosystem management runs strongly through the Eastside Project and its interim PACFISH strategy to protect anadromous fish habitat on public lands east of the Cascades, as well as the Clinton Forest Plan for managing federal lands west of the Cascades.

PACFISH is essentially a watershed management program for federal lands that provides habitat for Pacific anadromous fish. The program sets riparian goals, as well as riparian management objectives, standards, and guidelines. PACFISH also identifies key riparian management areas and watersheds, and calls for watershed analysis and restoration. Because it applies only to federal lands, however, even PACFISH stops short of truly comprehensive watershed management. Although the Clinton Forest Plan looks broadly at a huge geographic area extending from the Canadian border to San Francisco Bay, it emphasizes the protection of land within certain key watersheds. These key watersheds were selected either for contributing to anadromous salmonid and bull trout conservation or because they are sources of high quality water. The Forest Plan aims to protect these areas through actions based on watershed analysis—"a systematic procedure for charaCterizing watershed and ecological processes to meet specific management and social objectives."

National Marine Fisheries Service (NMFS): NMFS released its Proposed Recovery Plan for the endangered Snake River salmon in March 1995. Like the Forest Plan, the Proposed Recovery Plan aims to restore the health of an entire ecosystem, rather than focus on particular endangered species:

The goal of the Proposed Recovery Plan is to restore the health of the Columbia and Snake River ecosystem and to recover listed Snake River salmon stocks. Many of the recommended actions will directly benefit other species such as other salmon stocks, sturgeon, and bull trout. Implementation of the Proposed Recovery Plan should also conserve biodiversity, a factor that is essential to ecosystem integrity and stability.

The Proposed Recovery Plan is wide-ranging, addressing the tributary, mainstem, and estuarine ecosystems; harvest management; and artificial propagation. In evaluating the problems of the tributary ecosystem, NMFS recognized that many factors have played a role in the decline of the Snake River salmon: "Land and water management actions, including water withdrawals, unscreened water diversions, stream channelization, road construction, timber harvest, livestock grazing, mining, and outdoor recreation have degraded important salmon spawning and rearing habitats." The Proposed Recovery Plan's prescription for measures to address these problems is correspondingly comprehensive.

To address problems of salmon habitat in Columbia-Snake tributaries, NMFS concluded that "[a]n ecosystem-based approach that considers en tire watersheds and river subbasins is needed. Such an approach will en sure that all the physical, biological, and chemical processes and conditions that contribute to the development of productive salmon habitat are maintained." The Proposed Recovery Plan stressed that management planning is needed at both the ecosystem (Columbia Basin) scale and the watershed or sub-basin scale, and that ecosystem and watershed planning must be integrated.

The Proposed Recovery Plan recognized ongoing watershed planning efforts in several river basins and noted that they "provide useful templates for other watersheds." The basins identified are the Grande Ronde, Asotin Creek, Lemhi, Upper Salmon, and Tucannon. The plan also cited the joint salmon recovery effort of Wallowa County and the Nez Perce Tribe as "an example of a productive grass-roots process."

Regional Efforts (Northwest Power Planning Council (NPPC))

In December 1994, NPPC issued its revised Fish and Wildlife Program for the Columbia Basin. Like the NMFS Proposed Recovery Plan, the Fish and Wildlife Program analysed the causes of the regional salmon crisis and recommended corrective measures. And like NMFS, the Council attempted to approach salmon recovery on an ecosystem basis.

The Fish and Wildlife Program's prescription for salmon recovery addresses dam and river operations, harvest, and hatcheries, as well as habitat. Although the Fish and Wildlife Program is regional in scope, its approach to habitat restoration begins with "a `subregional' process that brings relevant interests together to address the needs of weak fish populations in particular watersheds." Here again, the ecosystem management strategy relies heavily on involvement at the local watershed level.

In fact, the Council's approach to habitat protection is based largely on voluntary and cooperative watershed, planning efforts. The Fish and Wildlife Program sets forth "habitat objectives" applicable to all watersheds but notes that these objectives are addressed principally to public lands within watersheds. However, the Fish and Wildlife Program also recognizes the necessity of habitat conservation efforts on private lands and calls for public and private parties to voluntarily join together in developing and implementing measures to restore fish habitat. The Council also specified that it expects progress toward habitat objectives to be achieved through watershed assessment, watershed management, collaboration, and locally adopted watershed plans.

The Council actually went a step further, making cooperative, watershed-based habitat protection efforts a specific element of the Program. The Fish and Wildlife Program calls for state and federal agencies, along with the Bonneville Power Administration, to support local watershed efforts. It also recommends that a model watershed project be implemented in each of the four major Columbia Basin states, based on the Grande Ronde, Upper Salmon and Lemhi Model Watershed, which had already been established under the Program.

Oregon's Watershed Programs and Other State Initiatives

Oregon has the oldest and most extensive statutory program relating to watersheds of the four major Columbia Basin states. Oregon established the Watershed Enhancement Program in 1987, and in recent years, the state's efforts to restore and protect watersheds have evolved and expanded. The Watershed Enhancement Program is administered by the Governor's Watershed Enhancement Board (GWEB), an eleven-member panel comprising the heads of five state natural resource commissions and boards and six state and federal officials. The Watershed Enhancement Program helps promote the state's goal to "enhance Oregon's waters through the management of riparian and associated upland areas of watersheds in order to improve water quality and quantity for all beneficial purposes." By 1995, GWEB had awarded nearly $2.3 million in grants for education and demonstration projects involving vegetation planting, fencing, and other watershed restoration measures.

Oregon's second program, the Watershed Health Program, was established in 1993 by House Bill 2215. This Bill directed the Strategic Water Management Group (SWMG), a state interagency council, to initiate a watershed management program based on a report that had already been issued by a public-private working group. The report recommended a wide variety of measures whereby the state could restore and protect watershed health.

House Bill 2215 also encouraged the establishment of voluntary local watershed councils as described in the working group report. The report envisioned watershed councils that would "represent a balance of interested and affected persons within the watershed" and facilitate citizen involvement in developing and implementing a watershed program. The report identified seven functions of these local watershed councils; these functions related more to communication, cooperation, conflict resolution, and public involvement than to actual resource management.

House Bill 2215 authorized SWMG to work cooperatively with any local watershed council, but provided that requests for state assistance "shall be evaluated on the basis of whether the requesting organization reflects the interests of the affected watershed and the potential to protect and enhance the quality of the watershed in question." As of April 1995, no fewer than thirty-six local watershed councils had formed in Oregon. Some of these councils were just getting organized, while others had produced plans and started to implement them.

The 1995 Oregon legislature passed a law that gave GWEB control over the Watershed Health Program. Because of the law, the Watershed Health Program has simply become a state program to support local watershed councils. The additional measures contemplated by House Bill 2215, most of which had not been implemented, were jettisoned from the law. The 1995 law provided some general guidance on the makeup of local watershed councils, the plans to be developed by these councils, and state priorities in promoting watershed health.It also revised the Watershed Enhancement Program to be more favourable to watershed councils. Thus, Oregon's watershed programs have been unified under GWEB, whose main role will be funding specific projects identified by local councils.

The 1993 legislature appropriated $10.2 million to fund the Watershed Health Program for two years, the bulk of which was to be spent for the South Coast/Rogue and the Grande Ronde Basins. As of early 1995, over one hundred individual projects in these basins had been completed or approved by SWMG for Watershed Health Program funding. The 1995 legislature allocated approximately $2.7 million in lottery funds to GWEB. The Washington Legislature passed a law in 1994 to establish a watershed planning program, somewhat similar to Oregon's, based on state agency coordination and project funding assistance. The statute made no changes in substantive law but established a Watershed Coordinating Council comprising representatives of ten state natural resource agencies. The Council was to coordinate watershed planning actions among state agencies, as well as with federal, tribal, and local governments. Both the statute and a later executive order directed the Council to prepare a wide-ranging report on watershed planning and restoration matters. In addition, the legislature appropriated $10 million for watershed restoration projects to be selected by state agencies.

The other two Columbia Basin states have not gone so far as to establish a statewide watershed management framework. (All four states do authorize water resource planning on a watershed basis, but most of these efforts do not involve comprehensive watershed management.) Both Idaho and Montana, however, have legislatively approved a watershed planning/management approach in particular river basins: the Henry's Fork in Idaho and the Upper Clark Fork in Montana. In both cases, however, the legislatures essentially approved councils that had already formed within these basins.

Local Efforts

Although government agencies at every level are now espousing watershed management principles, much of the activity in forming watershed councils and adopting watershed plans is coming from the local level—from within the river basins themselves. Many such efforts have been undertaken in the Northwest, each with its own story, and this Chapter merely samples them. The following pages identify one locally based watershed protection initiative in each of the four major Columbia Basin states and briefly explain their background, membership, and mission.

Idaho—The Henry's Fork Basin : The Henry's Fork of the Snake River drains a large area of eastern Idaho and a portion of northwest Wyoming. The Henry's Fork is a world famous trout stream; recreation and tourism are important to the Basin. Over 320,000 acres of cropland are irrigated from surface and ground water sources within the Basin. The timber industry is also economically important to the area.

The Henry's Fork Basin has long been the site of disputes between instream and out-of-stream water interests. The watershed council was initiated when a local environmental group, the Henry's Fork Foundation, approached the Fremont-Madison Irrigation District and asked the irrigators to support a local watershed council. The district not only agreed to participate in the council but also agreed to serve with the Henry's Fork Foundation as its co-facilitators. These groups formed the core of the council, which was formally recognized by the Idaho Legislature in 1994.

The Henry's Fork Watershed Council chose to be very inclusive in its membership:

The Council shall be comprised of citizens, scientists and agency representatives who reside, recreate, make a living and/or have legal responsibilities in the Basin, thus ensuring a more collaborative approach to resource decision making. The Council shall not be limited in the number of participants, with members organized into the following three component groups:

(a) Citizens Advisory Group—Members of the public with commodity, conservation and/or community development interests....

(b) Technical Team—The team shall be composed of scientists and technicians from government, academia and the private sector....

(c) Agency Roundtable—The Roundtable shall have representatives of all local, state and federal entities with rights or responsibilities in the Basin, including the Shoshone-Bannock Tribes.

The Council explained its approach as follows; As interests in the basin have diversified, the Henry's Fork has sustained increasing pressure to satisfy irrigation demand, hydropower requirements and instream flow needs for fisheries and recreation. These issues are the focal points of the Henry's Fork Basin Plan, passed by the 1993 Idaho Legislature.... In order to implement the recommendations and achieve long-term goals in the basin, an innovative, consensus-building process is needed which includes all resources and people in the watershed....... Citizens and agencies are beginning to recognize the importance of working together, as a rural community, to resolve the ecological problems in the watershed and to work towards a sustainable future for all concerned.

The council sees itself not simply as a planning body but as the implementer of management plans for the Henry's Fork Watershed. Perhaps the most significant of the plans is the Comprehensive State Water Plan for the Henry's Fork Basin, approved by the Idaho Water Resources Board in 1992 and the Idaho State Legislature in 1993. The council's support could advance several of the recommendations and goals of the Henry's Fork Plan, such as improved water conservation and increased screening of irrigation diversions.

Montana—the Upper Clark Fork Basin

Arising in western Montana and flowing northwesterly into Idaho's Pend Oreille Lake, the Clark Fork River has been called "Montana's largest and perhaps most abused river." The Clark Fork Basin is extremely scenic, and the mainstem Clark Fork is potentially an outstanding river for trout fishing. However, dewatering of streams by irrigation diversions is a substantial and chronic problem. The upper basin also has a serious water quality problem due to toxic metals, with four mining-related Superfund sites in the area.

Cooperative local watershed efforts in the Clark Fork Basin grew out of conflicts over water use, especially instream flows. In 1985, the Montana Department of Fish, Wildlife, and Parks filed a request to reserve water for instream flows in the Upper Clark Fork; shortly thereafter, a local conservation district filed a competing reservation request for irrigation water storage projects. As the dispute came to a head, the nonprofit Northern Lights Institute convened the principals in the Upper Clark Fork to discuss how to resolve the issues. Prior to the 1991 legislative session, the irrigators agreed to a moratorium on new surface water rights from the Clark Fork, pending further talks with the other interests. The 1991 Montana Legislature then chartered the Upper Clark Fork River Basin Steering Committee.

The 1991 legislation directed the Montana Department of Natural Resources Conservation to appoint the members of the steering committee, based on "their knowledge of water use, water management, fish, wildlife, recreation, water quality, and water conservation." In practice, the state selected many of the same people who had been involved in the Northern Lights negotiations. The statute required that the committee must at least include representatives of agricultural organizations, conservation districts, departments of state government, environmental organizations, industries, local governments, reservation applicants, utilities, and water user organizations.

The 1991 statute directed the Steering Committee to complete an Upper Clark Fork River Basin comprehensive management plan by the end of 1994. The plan was to consider and balance all beneficial uses of water in the Basin and make recommendations on how to resolve the Basin's water issues. The Steering Committee's plan, completed in December 1994, recommended several measures, including closure of the Upper Clark Fork to all new water rights, investigations of structural and nonstructural storage prospects, measures to "encourage" better water quality protection, and a pilot program to lease water rights for instream uses. The Montana Legislature codified many of the plan's elements in 1995. The 1995 statute also continued the Steering Committee, directing it to review the progress of management actions, make recommendations to the Montana Legislature, and serve as coordinator and facilitator on water issues in the Upper Clark Fork Basin.

Oregon-wallowa County

Wallowa County is the northeastern most county in Oregon. The major rivers draining the county are the Grande Ronde and Imnaha, both tributaries of the Snake. Most of the county is federal land, either within the Wallowa-Whitman National Forest or the Hells Canyon National Recreation Area. Timber, grazing, and tourism are key elements of the county's economy.

Declining salmon runs in Wallowa County and the imminent listing of Snake River chinook salmon under the Endangered Species Act caused Wallowa County and the Nez Perce Tribe to join forces in developing a salmon recovery plan. Wallowa County citizens expressed interest in fish restoration and the protections of existing economic uses of public lands. The Nez Perce Tribe wanted to rebuild the salmon runs, because, though it has its reservation in Idaho, the tribe holds treaty fishing rights in Wallowa County under its 1855 treaty with the United States.

The county and the Tribe formed a Wallowa County Salmon Recovery Strategy Committee. The Committee comprised Wallowa County citizens (representing grazing, timber, business, community, and environmental interests); representatives of the Nez Perce Tribe; and state and federal natural agency officials (from BLM, the U.S. Forest Service, and Oregon Department of Fish and Wildlife). In part, members were selected for their ability to work cooperatively on the issues; so-called "extremists" from the local community were not invited.

The Committee's mission was:

> *To develop a management plan to assure that watershed conditions in Wallowa County provide the spawning, rearing, and migration habitat required to assist in the recovery of Snake River salmonids by protecting and enhancing conditions as needed. The plan will provide the best watershed conditions available consistent with the needs of the people of Wallowa County, the Nez Perce Tribe, and the rest of the United States....*

In August 1993, the Committee produced the Wallowa County/Nez Perce Tribe Salmon Recovery Plan. The plan described salmon habitat requirements and desired conditions, analysed habitat problems, and identified possible solutions within twenty-seven Wallowa County sub-basins. The major focus was on forestland management, although the plan also addressed factors such as irrigation diversions and recreational river uses. The plan recommended implementation through a cooperative, consensus approach, even for activities involving federal lands. After producing the plan, the county formed the Wallowa County Watershed Enhancement Committee to provide for its implementation.

The plan already has achieved some of its goals by receiving favourable, albeit brief, mention in both the NMFS, Proposed Recovery Plan and the NPPC Fish and Wildlife Program. Moreover, the larger Grande Ronde Model Watershed Program later incorporated the Wallowa County/Nez Perce Plan.

Washington-the Yakima River Basin

The Yakima River drains much of south-central Washington before emptying into the Columbia River at the hi Cities. The upper Basin is largely national forest land, and the Yakama Indian Reservation occupies much of the Basin west of the River and below the city of Yakima. Irrigation is the region's dominant use of water and has helped the Yakima Basin become an enormously productive agricultural region.

Yakima Basin citizens launched a watershed initiative in response to a persistent drought that contributed to significant water shortages in the 1990s. The impetus came from a small group of agricultural producers and processors, area businesspersons, and concerned individuals. At first the group was primarily interested in advocating a project to pump Columbia River water up into the Yakima Basin. In March 1994, however, the citizens formed a more diverse Yakima River Watershed Council "to integrate a broad spectrum of water-based interests in the three-county area, encompassed within the Yakima River Basin, into a consensus process for the purpose of sustaining a sufficient water supply for all stake holders into the 21st Century."

The Yakima River Watershed Council claims a large and varied membership of individuals and a wide range of public and private entities. Its board of directors numbers an astonishing fifty people, and it has selected a twelve-member executive committee. Both the board and the committee include representatives of varied interests. In addition, the Council has set up technical committees to address a number of specific topics.

The Council's mission statement is "[t]o develop and implement, through consensus, a plan to provide consistent and adequate water to meet all economic, cultural, and natural environmental needs in the Yakima River Basin." The Council seems to have identified a number of possible options for meeting this goal, such as conservation, storage, and watershed restoration, but it is initially interested in acquiring streamflow data.

Tribal Efforts

Indian tribes are involved in a number of watershed management activities in the Northwest. Tribes participate in watershed efforts at many levels and in many capacities. Tribal activities range from directly implementing watershed restoration measures on and off their reservations, to serving on local watershed councils, to participating in federal ecosystem management efforts. The actions of the Confederated Tribes of the Umatilla Indian Reservation (Umatilla Tribes) illustrate the diversity of tribal involvement in watershed management. The Umatilla Tribes' reservation straddles the upper Umatilla River east of Pendleton, Oregon. The Tribes accepted this reservation in an 1855 treaty with the United States, under which they ceded over six million acres in present-day eastern Oregon and Washington. They retained hunting and fishing rights across this vast area, however, and thus have a major interest in several watersheds outside the Umatilla Basin.

The Umatilla Tribes have taken the lead in an effort to restore health to the Wildhorse Creek watershed, part of which is within their reservation. After studies showed that the Wildhorse Creek Basin created serious nonpoint source pollution problems for the Umatilla River, the Umatilla Tribes made it a major focus of their watershed restoration efforts. They met with the public and local landowners and expressed an interest in working on watershed projects in the Wildhorse Creek Basin. The tribal effort gained momentum after implementation of a demonstration project involving a major riparian landowner. The Tribes are currently undertaking several additional projects involving measures such as fencing, check dams, and streambank planting. Public agencies and private conservation groups have participated actively in these projects. The goal of the tribal effort is improved water quality and, over the longer term, restoration of salmon and steelhead to Wildhorse Creek.

Representatives of the Umatilla Tribes serve on both the Umatilla and Walla Walla Basin watershed councils. An Umatilla Tribal representative was also a charter member of the Grande Ronde Model Watershed Board. This representative explained the Umatilla Tribes are participating in the Grande Ronde program because it is an opportunity to work with people truly affected by issues and plans....

Tribal involvement benefits the [program] by bringing the clout of tribal government, which often works at the federal level, to a local effort.... Our hope for the Model Watershed... is that it will result in a restoration of the salmon economy for the tribes and that it will mesh in a positive way with the non-tribal economy in the basin.

Because of their interest in the overall health of salmon stocks in the interior Columbia Basin, the Umatilla Tribes have also taken a basinwide approach to water and watershed management. Their 1995 Columbia Basin Salmon Policy addressed the broad issues of water, harvest, and hatchery fish supplementation on a regional basis and called for actions to restore the health of the tributary, Columbia-Snake mainstem, estuary, and ocean ecosystems. The plan stated the Umatilla Tribes' support of certain measures to restore adequate instream flows and rebuild salmon habitat throughout the Columbia Basin.

The work of the Umatilla Tribes illustrates the broad range of tribal involvement in Northwest watershed issues but by no means tells the whole story. Tribes have played a key role in developing watershed strategies in several areas, including the Nez Perce Tribe's partnership with Wallowa County in preparing a salmon recovery plan and the Jamestown S'Klallam Tribe's role as coordinator of the Dungeness-Quilcene watershed planning effort. Northwest tribes have also worked collectively to address regional salmon recovery and ecosystem health issues through the Columbia River Inter-Tribal Fish Commission and the Columbia Basin Fish and Wildlife Authority.

The foregoing review of selected watershed management activities shows how popular this approach has recently become in the Northwest. Nearly all of these efforts have been launched since 1990, so they have not yet had time to produce major tangible results. Because they do not have a proven record of success, watershed approaches must be gaining acceptance for other reasons. The following section explores what these reasons might be.

Reasons for the Popularity of Watershed Approaches

The 1990s have been a time of strife in natural resources management in the West. Conflict has become commonplace, while consensus has proved elusive. "Ecosystem management" has itself been criticized, both by resource users who fear additional restrictions to protect natural resources and by environmentalists who oppose any weakening of current measures. Federal efforts at ecosystem management have been viewed with particular suspicion, partly because neither side trusts the federal agencies and partly because no one is certain what ecosystem management actually means.

The watershed movement, however, seems to have built a diverse constituency. The watershed approach has been initiated by federal agencies, rural counties, environmental groups, and business interests alike. Watershed protection seems to be one of the few concepts that currently enjoys support from many environmentalists, resource users, and management agencies. Of course, these groups do not necessarily support the idea for the same reasons.

Ecological Benefits : Watershed management offers the possibility that natural resources will be protected more effectively and comprehensively than they have in the past. Under a truly holistic approach to watersheds, individual resources and activities are not viewed in isolation. This approach acknowledges and understands the interconnectedness of everything in the watershed, all the lands, waters, and creatures, human and otherwise. The holistic approach entails recognizing all the effects resulting from humanity's actions and making decisions that protect and promote the long-term health of the watershed and all its inhabitants. In theory, at least, watershed management is truly a better approach.

The crash of Northwest salmon and steelhead runs has given the watershed approach a strong push. Not only has the salmon crisis made it necessary to do something, but it has shed light on the many factors causing the decline of freshwater salmon habitat. As stated in the NMFS Proposed Recovery Plan, "land and water management actions, including water withdrawals, unscreened water diversions, stream channelization, road construction, timber harvest, livestock grazing, mining, and outdoor recreation have degraded important salmon spawning and rearing habitats." Thus, federal agencies recognize that a wide variety of human activities adversely affect fish populations and habitat. Salmon advocates are hoping that watershed management can provide a stronger and more comprehensive approach to these varied threats, on private as well as public land.

Local Control : A second reason for the popularity of the watershed approach has more to do with economics and politics than with ecology. Many rural communities in the Northwest rely heavily on the use of public resources such as water and federal lands as their economic foundation. Such communities often believe they have little control over how these resources are managed, especially given recent injunctions against logging, grazing, and other activities on federal lands. Some of these communities have launched watershed councils, motivated largely by a desire to maintain existing economic activities and increase local control over public resources.

In eastern Oregon, at least, the potential economic impacts of the salmon crisis seem to have driven local watershed efforts. This was certainly true of the Wallowa County/Nez Perce Salmon Recovery Plan. The 1993 Oregon Legislature chose the South Coast/Rogue and Grande Ronde Basins to become the state's model watersheds-and thus to receive the bulk of $10 million in watershed program funding-because of concern that salmonid species in those basins would soon be listed under the Endangered Species Act.

The membership composition of some Oregon watershed councils strongly indicates that their primary goal is the protection of existing economic activities. For example, the Umatilla Basin Watershed Council was originally formed with thirteen members, of which eight to ten were closely associated with irrigated agriculture. Moreover, the original representative of "fisheries" interests on the Council was president of the Eastern Oregon Irrigation Association. For some watershed council members, restoring ecological health seems to be a secondary goal. As one member of the Grande Ronde board put it:

> *I represent private landowners on the Grande Ronde Model Watershed Board.... Our goal is better use of water. We recognize that fish survival is a major priority, but while establishing habitat for fish, we do not want to deteriorate the economic benefits that are based on water use. We see a need for timber production, agricultural irrigation, and flood control. Hopefully we can do these things, and improve fish habitat. In Wallowa County, resource users played a major role in developing a salmon recovery plan. The Wallowa County Salmon Recovery Strategy Committee was heavily weighted toward grazing and especially logging, with eight of eighteen members having direct timber ties. Forest management was a major focus of the Wallowa County Plan, and several of the plan's recommended management approaches could encourage logging. In commenting on a draft of the plan, both the Forest Service and the Bureau of Reclamation criticized the plan's emphasis on thinning dense stands of trees and indicated that this action could do more harm than good.*

An additional motivation for some local watershed efforts seems to be a desire to reduce the influence of "outsiders," such as government agencies, nonresident environmentalists, and federal courts. This was certainly a strong factor in the formation of the Umatilla Basin Watershed Council: "As we move forward we seize the opportunity to control our own destiny. We will be avoiding the potential for outsiders to demand action. We can identify our own solutions to the local problems."

Cooperation and Consensus

Many watershed management proposals provide that problems should be approached cooperatively and decisions be made by consensus. NPPC, in prescribing a coordinated approach to habitat planning, "urges all parties to undertake, collectively and voluntarily, the habitat assessment and restoration actions needed" for salmon recovery on a "truly collaborative" basis. For many local watershed councils, consensus is a bedrock principle. For example, the Idaho Legislature recognized and commended the Henry's Fork Watershed Council for its "nonadversarial, consensus-based approach to problem solving and conflict resolution among citizens, scientists and agencies with varied perspectives."

The consensus approach appeals to many people who believe that natural resource decisions are best made through a process of cooperation rather than conflict. Despite the increasing polarization over natural resources in the West, or perhaps because of it, these people call on all sides to come together and be reasonable. They maintain that if all sides have access to good, scientifically sound information, even traditional opponents can find common ground. That view may be overly optimistic, but many agencies nonetheless have embraced consensus-based watershed strategies.

At least two factors help explain this result. First, the agencies see voluntary, consensus approaches as the best way to gain the cooperation of private landowners and water users. Public lands cannot bear the entire burden of habitat restoration and species recovery, and the agencies believe it is better to address private property through cooperation rather than regulation. Second, many natural resource agencies seem eager to become less controversial. Battered by strong, conservative political winds, the agencies have sought to lower their profile by pursuing consensus at the local level.

Local watershed councils may have other reasons for operating by consensus. They may consider it the best means to promote watershed health. However, from the standpoint of conservation, the consensus approach has both benefits and drawbacks. The benefit of consensus is that it can reduce resistance and increase cooperation regarding management actions. The Wallowa County Plan states that nothing will work without consensus: "Any plan that does not have local ownership and support as well as the cooperation of government agencies will not succeed." On the other hand, consensus can certainly lead to paralysis or to a result that constitutes the "lowest common denominator." Local councils may also rely on consensus for economic reasons. Where a major goal of these councils is to protect existing economic uses of land and water, a consensus approach ensures that these uses will not be threatened by council actions. This is especially true where council membership is heavily weighted toward economic interests, and where community sentiment strongly favours existing water and land use activities. The major reason many people want consensus at the local level, however, undoubtedly has to do with the nature of rural communities.

People within these communities often have close ties to the place they live and to the other people who live there. They see firsthand the environmental, economic, and social effects of natural resource uses and policies. For better or worse, people in rural communities are constantly confronted with the human and ecological consequences of their actions. Thus, it is not surprising that many rural communities seek to sustain their watersheds in a way that works for the local population. People have to live there. In sum, watershed strategies have become popular in the Northwest for a variety of reasons. No one can yet say how effective they will be, because the phenomenon is too recent to have produced many tangible results. But even today, many watershed approaches show a basic, and perhaps fatal, flaw: not enough attention is given to water quantity-to restoring and protecting instream flows.

The Role of Streamflow Protection in Watershed Management

A healthy watershed requires some water to flow in its rivers. Many Northwest rivers, however, are completely dried up because water is withdrawn for consumptive uses, particularly irrigated agriculture. Watershed management strategies recognize this problem, but for various reasons they tend to concentrate more on improving land use practices and water quality. The four major Columbia Basin states all provide some legal protections for instream flows, but these protections are neither adequate nor connected to watershed management programs. If such programs are to succeed in restoring Northwest watersheds to health, they must recognize instream needs and provide means to restore and protect streamflows.

Streamflows and Watershed Health

Largely because of water diversions for out-of-stream uses, many rivers and streams in the Columbia Basin dry up in the summer and many others are severely depleted. The diversion of entire rivers for private uses is wholly legal under the prior appropriation doctrine that dominates western state water law. The states issue permanent rights allowing a certain amount of water to be taken for a particular "beneficial use," such as irrigation or household use. The owner of such a water right may take as much water as the right allows, even if a river is drained dry as a result, unless that water is needed to satisfy an older, and thus superior, water right.

In identifying impediments to recovery of the endangered Snake River salmon, the National Marine Fisheries Service (NMFS) recognized low streamflows as a major problem and identified irrigation water withdrawals as the primary cause, especially in the Snake River Basin. Irrigation accounts for nearly eighty-eight percent of all water withdrawals in the Pacific Northwest. Irrigation water needs are greatest in the summer and early fall, during a time when streamflows are naturally at their lowest due to scarce rainfall and depleted snowpacks.

The impacts of agricultural water withdrawals in a particular watershed may be severe, whether irrigators primarily pump ground water, as in Idaho's Big Lost River Basin, or divert surface water, as in Oregon's Umatilla River Basin. The Oregon Water Resources Commission found that in the Umatilla Basin, "[l]ow streamflows are the chief limiting factor to salmonid production. Low streamflows impede and block fish migration, increase water temperatures, and contribute to reduced habitat and competition from warm water fish species."

Two recent studies underscore how vital instream flows are for salmon and steelhead. The Umatilla Tribes, who evaluated wild steelhead in the Umatilla River, and the Centre for the Study of the Environment, who evaluated spring chinook in the Rogue River, both have established a strong positive relationship between river flows and returns of anadromous fish two to four years later. The Rogue River study showed that adult salmon returns were more strongly related to minimum streamflows than any other factor studied, including harvest and hatchery production. It also suggested that dams and water withdrawals are major problems for salmon survival on the Rogue. Daniel Botkin, Director of the Centre for the Study of the Environment, concluded that the lesson of his study "is do everything you can to raise minimum flows."

Many resource management agencies and watershed protection strategies recognize the importance of instream flows and call for measures to improve them. For example, the Proposed Recovery Plan repeats the Snake River Salmon Recovery Team's recommendation that State fishery agencies ensure that actual streamflow withdrawals for diversions are not in excess of legal water rights. If insufficient streamflows are still occurring, reduction of irrigation withdrawals should be sought wherever such withdrawals lead to insufficient streamflow for habitat, impede downstream or upstream passage, or infringe upon water quality control at any downstream location.

Other management plans that recognize the need for instream flows to restore fish habitat and watershed health include the Northwest Power Planning

Council's Fish and Wildlife Program, the federal PACFISH strategy, the Grande Ronde Model Watershed Program's Operations-Action Plan, and the Umatilla Tribes' Salmon Policy.

Water Right Troubles

Despite recognizing the importance of instream flow restoration and protection, watershed management plans generally attempt to provide fish habitat by other means, such as improving land use practices and adding "structure" to streams. Sometimes this approach can be explained largely by agency jurisdiction: Because the U.s. Forest Service and U.S. Bureau of Land Management are land management agencies, their plans focus on land use. But other plans, even though they aspire to holistic watershed management and stress the need for water instream, do not address streamflows effectively. Instead, they are stymied by western water law and politics.

As an example of this, NPPC's Fish and Wildlife Program calls for a holistic, watershed-based approach to salmon habitat: "A total watershed perspective, in which fish needs, land and water conditions, and local, private and government initiatives are viewed together, will play an essential role in the ultimate success of efforts to rebuild salmon and steelhead runs." The Program specifically recognizes the need for instream flows, and calls on the states, the Bureau of Reclamation, and the Bonneville Power Administration to take certain steps to ensure adequate flows. But Congress has limited NPPC's ability to address streamflows, by requiring the interstate agency to defer to state laws governing water allocation:

Congress and the Council recognize that this program must be implemented within a complex scheme for allocating rights to use Columbia River Basin water. As noted in the Northwest Power Act,... nothing in this program authorizes appropriation of water, affects rights to water or jurisdictions over water, or establishes the respective rights to water of the federal government, individual states, Indian tribes or individuals. The Council assumes that the federal implementing agencies will work hard to develop cooperative and creative ways to implement the program's water flow measures with those requirements in mind.

Thus, NPPC maintains that although water is urgently needed for instream use, it generally cannot be provided except through "cooperative and creative" means. The Northwest Power Act's specific disclaimer on water rights illuminates an important point. Watershed management efforts throughout the region proceed on essentially the same basis: Go forth and do good, saving fish and promoting ecosystem health in a holistic manner, but Don't Mess with Water Rights. This result is determined in part by legal considerations. Even though water rights confer only a limited right to use water owned by the public, they are generally considered to be property rights. State law predominately determines how western waters are used. Even federal water laws, such as the Clean Water Act and the reclamation laws, acknowledge the primacy of state law in water allocation.

The biggest constraint on watershed management efforts to improve instream flows, however, may be politics. For the most part, Northwest water users, especially irrigators, have successfully resisted changes that could provide environmental benefits. Irrigators and other water right holders have enormous influence in the capitals of Northwest states. The legislatures of these states will go to great lengths to protect existing water uses and users. Water users' political influence is almost certain to insulate them from any watershed management strategy at the state level. At the local level, especially in rural areas heavily dependent on agriculture, irrigation and ranching interests have even greater control. Given that many watershed efforts seem to be motivated largely by various parties' interests in pursuing consensus, reducing public conflict over environmental issues, and asserting local control over natural resources, political constraints seem even more significant. No federal or state official who wants to avoid conflict is likely to make an issue of water rights. Even more clearly, no watershed council with a strong interest in protecting the local economy is likely to propose substantial changes in water use. It is easier for all concerned to focus on somewhat less controversial matters, such as installing fish screens, planting trees along riverbanks, and keeping cattle away from riparian areas. While land use changes may improve flows at certain times by helping to restore the natural hydrograph, such measures offer limited benefits on overappropriated rivers with inadequate streamflow protection.

Water Rights and Local Watershed Strategies

In some Northwest river basins where there is enough water to meet instream and out-of-stream demands, instream flow protection need not be the top priority of a watershed management approach. In many basins, however, seasonal low flows, exacerbated by out-of-stream diversions, are a major problem, and streamflow restoration is an urgent need. Unfortunately, watershed management efforts in such basins have been hindered by water rights concerns, both legal and political. The Lemhi River Basin is a case in point. The Lemhi flows into the Salmon River in east-central Idaho, and the Basin has been heavily developed for agricultural irrigation. As explained in a 1990 state/tribal review of the Basin, "the Lemhi's flow is totally appropriated for irrigation; the use of water from the watershed for irrigation influences discharge patterns more than any other factor." The review also noted that "irrigation diversions were impacting anadromous fish production in the Lemhi and the headwaters of the Salmon River as early as the 1850s" and that "[i]rrigation withdrawals have reduced flows, limiting juvenile and adult passage and increasing water temperatures, often to critical levels for anadromous fish during summer months."

In 1992, the Lemhi Basin was selected as a model watershed project under NPPC's Fish and Wildlife Program. A fifteen-member advisory committee was formed "representing local residents, Indian tribes and conservation groups," with the support of a technical committee of agency professionals. Like many local watershed councils, the Lemhi advisory committee developed a vision statement and is guiding the preparation of a watershed management plan. Moreover, the Lemhi watershed effort has produced some actual water instream at a critical time for salmon. Under a multi-party Memorandum of Understanding, irrigators jointly agreed to forego diversions for a twelve-hour period up to three times a year if low flows cause fish passage problems in a certain reach of the Lemhi. On July 21, 1994, operation of the "Lemhi fish flush" dramatically raised flows in the targeted reach, although it is not clear the operation actually helped any fish. The "flush," along with other steps taken or planned in the Lemhi Basin, show a serious commitment to improving fish habitat.

Nonetheless, low flows caused by irrigation withdrawals continue to be the major problem for salmon habitat in the Lemhi watershed. As stated in the NMFS Proposed Recovery Plan, even though the Lemhi process has been very helpful in building partnerships and has resulted in measurable subbasin improvement, difficulties still remain because: (1) The river is over-appropriated for irrigation, and (2) major tributaries are regularly dewatered for irrigation. No significant improvement in spring/summer chinook salmon passage and rearing habitat for the Lemhi River as a whole is likely to occur unless changes in local agricultural practices are made. In the Henry's Fork Basin, a cooperative process produced some worthy potential projects beneficial to streamflows, but legal issues involving water rights have impeded these projects. According to Jan Brown, cofacilitator of the Henry's Fork Watershed Council, "[w]e've got at least four projects that have received endorsements from the Council, where farmers want to return water to the stream, but they're running into their own water law as an obstacle." Idaho water law does not allow the lease or transfer of private water rights for instream use.

In Oregon, such leases are allowed by law but are sometimes hindered by politics. The Grande Ronde Model Watershed Board recently refused—officially, "deferred"—a funding endorsement for a project involving a short-term, voluntary lease of water rights for instream uses in the Lostine River Basin. One board member voiced opposition, as a matter of principle, to any separation of water rights from the appurtenant land. The Board voted 6-5 to defer the proposal. Thus, the Grande Ronde Board was split down the middle over the issue of a short-term voluntary lease of water rights for instream uses. This result illustrates the potential political problem that consensus-seeking local watershed councils may encounter as they attempt to do anything regarding existing water rights.

State Instream Flow and Water Planning Efforts

Idaho, Montana, Oregon, and Washington all have laws allowing state agencies to establish instream water rights or minimum streamflows. All but Idaho also allow private water rights to be converted to instream use, at least temporarily. In addition, each state has laws providing for a comprehensive state water plan to be developed watershed by watershed. For the most part, these streamflow protection and water planning programs are not directly connected with the watershed efforts described. Experimental planning efforts in Montana and Washington, however, produced locally driven watershed management plans that attempt to address instream flow issues at the basin level.

State Instream Flow Laws: State agencies have the primary role in establishing water rights for instream uses in the Northwest states. Idaho allows its Water ResourceBoard and Department of Water Resources to set minimum streamflows, subject to disapproval by the legislature. Montana allows both state and federal agencies, along with political subdivisions of the state, to apply to reserve water "to maintain a minimum flow, level, or quality of water" in any of six designated river basins. Oregon authorizes three state agencies to seek instream water rights for fish and wildlife, water quality, recreation, and scenic attraction. Washington grants its Department of Ecology the exclusive authority to establish minimum water flows or levels to protect the public interest. Oregon was the first state in the region to pass its original minimum streamflow law in 1955; Montana was last, enacting its first general law on instream flows in 1973.

A basic problem with these state instream flow programs is their late inception. Many Northwest rivers were fully appropriated by the early 1900s; on such rivers, late twentieth-century water rights often receive no water, especially in the irrigation season. A junior instream water right on an over-appropriated river does have some benefits; it can keep the shortage from getting worse and may block certain transfers that could harm streamflows. But it does not stop senior water right holders from drying up a river. In other words, instream water rights can protect existing streamflows but cannot necessarily restore them.

Recognizing this shortcoming, three Northwest states have enacted laws allowing private water rights to be converted to instream use under certain conditions. Oregon's 1987 Instream Water Rights Act allows any person to purchase, lease, or accept a gift of an existing water right and convert it to an instream water right, either permanently or temporarily. Washington law authorizes its Department of Ecology to acquire water rights in certain river basins and convert them to instream use under a "trust water rights" program. The 1995 Montana Legislature established two programs—one applying statewide, the other applying only to the Upper Clark Fork Basin—that allow a temporary change of water right for instream use "to benefit the fishery resource."

Oregon and Washington have also provided a legal mechanism to allow a certain portion of water "saved" through water conservation measures to be dedicated for instream uses. The Oregon Conserved Water statute(2070 allots at least twenty-five percent of saved water to instream uses, and the percentage may be higher if public money finances more than twenty-five percent of the conservation measures. Washington's trust water rights program provides that "net water savings" from conservation projects be treated as trust water rights, although these may be used for either instream or consumptive uses. The rationale for these programs is as follows:

Historically, appropriated water which was conserved through efficiency or not beneficially used was considered reduced waste and returned to the stream for use by other water right holders or reappropriation by the state. No means existed for redirecting the use of this water. Water not put to beneficial use was considered forfeited. The trust water rights program allows water saved through efficiency to be transferred to a new use, while retaining the original priority date.

Aside from a few recent, short-term water right leases in Oregon, however, little use has been made of these innovative programs to restore instream flows. Explanations for this lack of activity vary. Water users interested in making water available for instream uses may face uncertain results, indeterminate transaction costs, a potential reduction in property values resulting from an apparently smaller water right, and almost certain peer pressure from people and groups who believe that water should not be separated from the land. Incentives for instream water transfers, leases, and conservation projects have not yet overcome such obstacles, particularly because water users have very little to lose by maintaining the status quo.

State Water Planning Programs

All four Northwest states have laws requiring comprehensive state water resource planning. In each case, the plans must be adopted by the state water resource agency, board, or commission. Each plan involves a mix of local public and state government involvement, with Washington's process being the most locally driven. Of the four statutes, Idaho's is the most detailed and prescriptive, while Washington's is the least so. All are essentially similar, involving several common elements:

- Information gathering. Idaho, Montana, and Oregon all require that the responsible agency conduct certain studies as the basis for planning.
- Comprehensiveness. The four statutes call for integrated and coordinated, comprehensive plans that address the use of all of the state's water resources.
- Maximum development/use. All statutes but Washington's emphasize the desirability of maximizing the amount of water available for use by the people of the state.
- Planning by basin. All four statutes contemplate that the state water plan will be developed at the basin or watershed level.
- Public participation. Three statutes require a local public hearing before a basin plan is adopted, while the Washington statute provides for more general public participation.

Despite these basic similarities in their statutes, the Northwest states produced four very different types of water plans. One major distinction between the four state water plans is their treatment of instream flows. The following subsections review selected water planning materials from each of the four states.

Idaho : The current Idaho state water planning statute was adopted in 1988. In 1978 the Idaho Legislature adopted the Water Resource Board's original water plan but specified certain "changes" that established state water policy on thirty-seven subjects. These subjects ranged from identifying unrecorded water rights, to protecting potential reservoir sites, to monitoring radioactive waste disposal. Policy No. 1 is "'protection of existing water rights." Policy No. 6 states: "Water rights should be granted for instream flow purposes. The legislation authorizing this policy should recognize and protect existing water rights...." Policy No. 32 sets the famous target flow of zero at Milner Dam on the Snake River.

Under the 1988 law, the Comprehensive State Water Plan is to be written at the watershed or sub-basin level. The Idaho Water Resources Board is responsible for the plan, although local input weighs heavily in plan development. These plans, not unlike the planning requirements under the National Environmental Policy Act, consider the environmental conditions within a basin, discuss water resource issues affecting the area, and analyse alternatives. In terms of "Actions and Recommendations," the plans focus heavily on designating particular stream reaches as natural or recreational. Within these reaches, the plans prohibit or limit such activities as the construction or expansion of dams, the construction of hydropower projects or diversion works, or the conduct of various activities within the stream channel.

With respect to instream flows, the Middle Snake Plan identified minimum streamflows that have been established or applied for under Idaho law. It noted that the State Water Plan specifies a minimum flow of zero at Milner Dam. The Middle Snake Plan also recognized, however, that the public interest requires increased flows in the Snake:

> *As a long-term goal, the Water Board will work toward higher river flows during the summer months. The Idaho Water Resource Board would like to see more than a zero flow at Milner and will continue to examine options to secure flow throughout the year at the dam or main powerhouse.*

Increased flows would improve some aspects of water quality and fish habitat, and restore some of the scenic beauty to Twin Falls, Shoshone Falls, and many of the smaller, less famous waterfalls within the reach. At this time, there is no ready mechanism to provide this water. Increased irrigation efficiency could lead to increased operational flexibility in the Snake River, and perhaps more water through the Middle Snake reach during low-flow periods....

But while the Middle Snake Plan acknowledged the need for greater streamflows, its only recommendation was that the state study methods for increasing flows in the Middle Snake reach. The need for instream flow protection was a recurring theme of the Henry's Fork Plan. That Plan noted that irrigation water diversions create low flow problems for fish in parts of the basin, recommended an examination of the need for minimum flows, and stated that such flows should be sought where a need is identified. The Plan also recommended that water of sufficient quantity and quality be protected for recreation. However, the Henry's Fork Plan also supported further economic development of the Basin's waters, particularly for irrigation. In general, the Plan placed greatest emphasis on new development projects; in assessing water supply and water conservation, the Plan devoted only four sentences to minimum streamflows but seven pages to potential new storage sites.

Montana: The Montana statute does not provide the Montana Department of Natural Resources and Conservation with much guidance, except that the sections of the state water plan are to be developed by river basin. Thus far, however, Montana's water planning has not proceeded that way. Instead, the state has produced the plan in sections addressing statewide issues, such as water storage, integrated water quality and quantity management, and drought management. These sections identify and briefly discuss issues, declare state water policy on these issues in one paragraph, and set forth options and recommendations for resolving the issues.

The subsection on instream flow protection recognizes that Montana must do more to protect streamflows effectively. It adopted the following policy statement:

> *Instream flows are an important use of water, and mechanisms should be developed and refined to protect and enhance instream resources. However, instream flow protection activities must not adversely affect existing water rights and should be weighed and balanced against alternative future uses of water.*

The plan makes four recommendations: 1) revise the water rights permitting process to improve instream flow protections; 2) consider making instream water reservations more secure; 3) allow voluntary water right leases for instream uses; and 4) conduct research on matters related to streamflows.

Montana now seems ready to move ahead with water planning at the river-basin level. The state intends to initiate pilot projects in one or two basins, beginning with the formation of basin advisory committees representing all affected water users and interests. The hope is that these local committees can address problems through a cooperative approach. The committees are to identify their basins' water resource issues, evaluate and select options to resolve them, and incorporate them into a basin plan. Such plans will be reviewed by the Department of Natural Resources and Conservation; upon approval, the plans will become part of the State Water Plan.

Local water planning efforts are proceeding in roughly a dozen Montana basins, with various levels of state involvement and assistance. These basins include the Flathead, Musselshell, Muddy, Bitterroot, and Kootenai. Of these basin planning efforts, the Flathead is perhaps furthest advanced. In one Montana watershed, a local committee has already developed a plan that has essentially received legislative approval.

The Upper Clark Fork River Basin Steering Committee produced a water management plan for that basin in December 1994. The plan described the steps that were taken in developing the plan, including the public involvement, and discussed existing basin conditions in some detail. It also made recommendations on nine matters relating to existing and new water rights, water management, water quality, and instream flows.

The Upper Clark Fork Plan acknowledged that irrigation creates a major problem of stream dewatering: Over 471 miles of streams in the Basin, including nearly 93 miles of the Upper Clark Fork mainstem, are listed by the state as "chronic[ally] dewatered." In response, the Plan recommended closing the Basin to most new waterlights, while protecting existing rights. It recommended a pilot program to allow the leasing of water rights for instream uses and suggested that the Montana Department of Fish Wildlife and Parks "continue to seek willing landowners to help solve dewatering problems."

Oregon: Oregon's approach to basin planning is fairly similar to Idaho's approach. Statutes establish the basic policies under which the state water resources program is to proceed. As in Idaho, the first policy is protection of existing water rights. Oregon law also favours "maintenance of minimum perennial stream flows sufficient to support aquatic life, to minimize pollution and to maintain recreation values... if existing rights and priorities under existing laws will permit."

The Oregon water resources program is a collection of eighteen basin plans or programs, each adopted for a particular watershed. Under these plans, the Water Resources Commission makes a series of findings regarding the water resources of the basin, existing instream and out-of-stream water uses, and possibilities for water demands and development in the future. The primary result of basin programs is that waters from a particular source are "classified" for particular purposes-that is, new uses of water are allowed only for the purposes specified in the basin program. For example, the South Coast Basin Program classified the waters of the West Fork Milicoma River and tributaries above Stall Falls for municipal, domestic and livestock uses, irrigation of lawns and noncommercial gardens not exceeding one-half acre in area and instream use for recreation, fish life, and wildlife.

Most or all of Oregon's minimum perennial streamflows were adopted as a part of basin programs. Minimum perennial streamflows are not water rights, but rather administrative rules protecting flows below specified levels, subject to certain exceptions. In contrast, instream water rights are water rights for a given flow in a given reach with a definite priority date. The Water Resources Commission converted the vast majority of minimum perennial streamflows into instream water rights, as directed by statute. The Water Resources Commission no longer sets new minimum perennial streamflows through the basin programs. Instead, it considers instream water right requests along with other water right applications.

Oregon's basin programs address streamflows in several ways. For example, the Grande Ronde River Basin Program found that instream flows in the basin were insufficient to meet instream needs, and that existing consumptive water uses could cause "zero flows or flows approaching the zero level during critical low flow periods." This program also concluded that the 1961 minimum perennial streamflow for the Grande Ronde River was inadequate and that a state agency's 1985 minimum perennial streamflow request was "more appropriate for the support of aquatic life" than a previous request.

In addition, the Grande Ronde Program established minimum perennial streamflows in various sub-basins and stated that new appropriations would not be allowed when flows fell below those levels. On the other hand, in the 1984 South Coast Basin Program, the Water Resources Commission rescinded two minimum perennial streamflows and reduced two others that had been established in the 1964 program, while retaining most of them and establishing two new ones. Finally, both programs classified waters only for specified uses, and did so partly to protect remaining streamflows from impairment caused by new appropriations for other uses.

Washington: While Washington has statutes on water resource planning, a document known as the Chelan Agreement primarily guides the state's planning actions. The purpose of the Agreement is to "establish procedures to cooperatively plan for the management of water resources in Washington State to best meet the goals and needs of all its citizens." The Chelan Agreement resulted from a 1990 conference involving over 150 people from throughout Washington, representing agriculture, business, the environment, fishing; and recreation interests, along with state, local, and tribal governments. A major impetus for the Chelan Agreement was the state's desire to resolve conflicts over tribal treaty fishing rights, as played out in the long-running U.S. v. Washington litigation.

The Chelan Agreement calls for cooperative water resource planning at the river-basin level. Eight interests, or "caucuses," must be invited to participate in the process: state, local, and tribal governments; agriculture; environment; sport/commercial fisheries; recreation; and business.266 Additional interests may gain "caucus" status by consensus of these groups or by petition to and upon approval by the Washington Department of Ecology (Ecology). Each caucus chooses its own representatives. One of the participating state, local, or tribal government entities plays a coordinating role. Interested federal agencies are to be "invited to participate in whatever manner is dictated by that region." The Chelan Agreement states a preference for consensus decisions, but also establishes how to proceed in the absence of consensus:

Each caucus will have one voice in decision-making. The planning group will attempt to reach consensus whenever possible. In cases where consensus is not possible, decisions will be made by a consensus of the government caucuses and a majority of the interest group caucuses. Minority reports, if prepared, shall be included in the plan document.

Where consensus among the governments (tribal, state, and local governments) and/or a majority of the interests is not achievable, the Department of Ecology shall assume the lead role in assuring that the plan is completed for the pilot projects in a timely fashion, not to exceed twenty-four months.

The regional groups are to prepare water resource management plans that must address, at a minimum: 1) ground water, 2) surface water, 3) consumptive needs, 4) nonconsumptive needs, and 5) the relationship between surface and ground water. Ecology must review completed plans for compliance with applicable federal and state laws and regulations; it can accept or remand the plans, but may not change them. Ecology and local governments must prepare and adopt any new regulations or ordinances needed to implement the plans.

The Chelan Agreement required Ecology to select at least two pilot projects for regional planning. Ecology selected the Methow Basin in north-central Washington and the Dungeness-Quilcene Basins of the Olympic Peninsula. Plans have now been prepared for these basins, and Ecology intends to proceed with rulemaking on both plans. At the time of this writing, the Washington Legislature appeared likely to appropriate funds requested for the implementation of these plans. The Dungeness-Quilcene Plan makes a variety of findings and recommendations, many of which deal directly with instream flows.

For example, the Plan recommends the setting of instream flows on certain rivers, the denial of new water rights in one area, and the use of water conservation and the trust water rights concept. Even more remarkable is that the Plan contains an agreement among the various users to share shortages during periods of low water:

The gap between the needs of the fish expressed by recommended instream flows, and the present instream flow after withdrawals for agriculture, municipal, business and future growth needs is substantial. This is amplified by the poor condition of fish habitat, the lack of conservation, the inefficiency of irrigation delivery systems in some areas, and the other uses which take water from the system. Under the gap strategy, the Regional Planning Group agrees to acknowledge that a discrepancy exists, is likely to continue indefinitely, and that to some extent the parties will have to live with it.

In this plan, the RPG makes recommendations intended to bring the sides of the gap closer together. Through participating in shared sacrifice, the members of the planning group have agreed to share the pain and share the gain. When the weather and other conditions provide abundant flows, ample water is available for all uses; when the opposite occurs, during times of low flows and critical needs for both fish and human uses, ail sides agree to restrict uses, and to share water equitably.

In other words, the Dungeness-Quilcene Plan essentially contains an agreement to manage the Basin's water resources based on equity and shared sacrifice, rather than water right priorities. The plan does not specify how the agreement was to be implemented. The Methow Plan is somewhat less revolutionary but still progressive. One of its six major conclusions provides that "[i]nstream flow must be increased to improve fish and wildlife habitat and preserve and enhance the unique quality of the Methow Valley while allowing for growth." New water uses must come from storage or from existing uses, through conservation, transfers, and agricultural improvements. The Methow Plan also produced an interdependent package of recommendations to address the Basin's water problems. These recommendations include setting efficiency standards for domestic uses and alfalfa and orchard irrigation;

establishing a "water bank" for water saved through conservation measures as the exclusive source of water for new uses; allocating saved water, with a ninety percent share for instream flows; targeting and increasing enforcement activities; and requiring measuring devices.

Streamflow Protection as a Component of Watershed Management

A wide variety of watershed and water management activities are going on throughout the Northwest. Looking broadly at these activities, one could conclude that a unified and comprehensive approach to water and watersheds is possible. The basic elements already exist in many places. Federal and regional agencies are focusing on watersheds, with the goal of ecosystem management. All four states have streamflow protection laws, and all have water resource planning at the basin level, at least on a pilot basis. Local watershed groups have formed to address problems in many river basins, and Indian Tribes are actively involved in watershed management efforts at all levels.

Figure: *Watershed Modeling for Stream Ecosystem Management.*

Some of the local watershed groups are beginning to deal with streamflow issues. A few of these efforts—as in the Lemhi, Upper Clark Fork, Methow, and Dungeness-Quilcene Basins—have produced some initially positive results, at least on paper. In other areas, such as the Henry's Fork and Grande Ronde Basins, a framework is now in place that could produce future benefits for instream flows. In those places where the cooperative local approach seems to be working best, a recurring theme is historic conflict. In many of these basins, years of discord over water issues finally brought people to the table. This is true at least in the Henry's Fork, Methow, and Upper Clark Fork Basins. Traditional adversaries such as irrigators and tribes, and developers and environmentalists, seem most likely to try cooperation when all else has failed, when all participants are sick of fighting, and when all sides have something to lose from the status quo.

Throughout the Northwest, water and salmon issues are highly controversial. Cooperation on these issues cannot always be expected, at least at the state and regional levels. For the reasons, however, the watershed approach seems likely to grow in popularity. A major unresolved question is whether these approaches will meaningfully address instream flows. The early indications from basins such as the Upper Clark Fork, Methow, and Lemhi offer some hope. And a few state and regional proposals recognize the need for watershed management to include instream flow restoration and protection.

Washington Governor Mike Lowry proposed legislation in 1995 to provide for genuinely comprehensive, statewide watershed management planning at the river basin level. This planning would be patterned after the successful Methow and Dungeness-Quilcene pilot projects, within the general framework of the Chelan Agreement. As stated in the governor's proposed legislation, such plans will resolve uncertainties about water quality, water resources, and habitat conservation measures necessary to preserve the environment, provide economic prosperity, and avoid federal or judicial interventions stemming from noncompliance with federal treaties and laws.......

It is further the intent to implement the plans through federal, state, local, and tribal authority, including the laws related to water quality, water allocation, public water supply, fish and wildlife protection, forest practices, and local land-use planning. Watershed resource management plans must therefore be comprehensive, addressing at a minimum water quality, ground and surface water availability and use, economic development, and habitat conservation. The plans would include recommendations on a wide variety of matters, including water use efficiency and conservation standards, water quality measures, instream flow requirements, metering and reporting requirements for new and existing water rights, conditions and limitations on new water rights, priorities for acquisition and use of trust water rights, and land use controls and incentives for habitat preservation and restoration. In Oregon, a diverse public and private working group issued a report in 1992 that laid the foundation for the state's Watershed Health Program. Subsequent legislation directed the state to initiate the program using the 1992 report as "a framework and guide." The primary focus of that legislation was establishing voluntary local watershed councils, which was one of the recommendations of the report. But the working group's report listed numerous other "proposed watershed management tools" and indicated that developing these tools was a high priority for successful implementation of a watershed management strategy.

Many of these tools could directly or indirectly benefit instream flows, including 1) improving water use efficiency, 2) allocating a percentage of transferred water for instream uses, 3) imposing a public interest test on transfers, 4) requiring measurement and reporting of water use, and 5) improving enforcement to benefit instream flows. These tools would bolster

Oregon's existing, relatively progressive laws regarding instream flows, such as those allowing instream leases and transfers of water rights and allocating a portion of conserved water for instream use.

The most ambitious proposal for comprehensive watershed management in the Northwest has come from Angus Duncan, former chair of the Northwest Power Planning Council (NPPC). In 1994, Duncan proposed a Columbia Basin Watershed Planning Council. Duncan suggested the NPPC "be specifically charged by the U.S. Congress with setting forth a general plan for the conservation and efficient use of the waters and lands affecting those waters of the Columbia-Snake River Basin." Duncan's proposal is truly far-reaching, involving Canadian and oceanic habitats as well as U.S. matters. It is equally expansive in the range of issues to be addressed: 1) water quality and water quantity, 2) consumptive and nonconsumptive water uses, 3) species conservation, 4) actions needed to achieve and maintain sustainability, 5) economic analysis, and 6) distribution of costs. These proposals have not fared well politically. Despite a cautiously worded letter of support from Senator Mark Hatfield (R-Or.), Duncan's proposal proved to be a nonstarter. Most of the new tools proposed in the 1992 working group report on watersheds are not under active discussion in Oregon. That report is no longer the basis of Oregon's watersheds program, which now consists almost entirely of supporting local watershed councils and their chosen projects. And Governor Lowry's proposal ran into trouble in a fractious Washington Legislature, which was unable to deal effectively with water issues.

These political results are not surprising, given the anti-government attitude now prevalent in the Northwest. Any proposal that raises the prospect of regulation, or even hints at an increased state or federal role in natural resource management, is likely to go nowhere in the current political climate. At present, the major political push is for watershed strategies that are locally controlled, strictly voluntary, and purely cooperative. Such local efforts are vitally important because they can promote public understanding and involvement in watershed matters, can increase conservation activities on private lands, and can establish working relationships among institutions and people who formerly did not talk-either out of ignorance or antipathy. Perhaps most importantly, local watershed efforts can help build support for meaningful action, cooperation, and change. Without local support, most resource users will continue to distrust anything labelled "environmental"; even solutions beneficial to all sides, such as water right leases for instream flows, will be difficult to sell. Equally important, however, these local initiatives must be linked with state, tribal and federal activities. Land management programs, water resource planning efforts, and instream flow laws all relate directly to watershed health. These activities should support local initiatives by providing information, resources, and a legal basis for management actions. Conversely, local efforts should support federal, state, and tribal goals and programs, particularly with respect to resources owned by the public such as water, fish and wildlife, and federal lands. Actions involving natural resources typically affect many interests-private resource users, local communities, Native Americans, and the public—and any watershed management effort must take account of all these interests, as well as the needs of future generations. Finally, it is crucial that voluntary, cooperative, local efforts do not supplant existing laws. Implementation and enforcement of federal, state, and tribal laws must not be sacrificed for the sake of consensus.

Watershed planning and management activities are moving ahead throughout the Northwest. Federal and regional agencies, state legislatures, Indian Tribes, local governments, and private citizens are embracing the watershed approach. The concept has grown in popularity for disparate and sometimes conflicting reasons. Agencies and conservationists perceive a need for a more comprehensive approach to resource management and species preservation. Local communities and resource users desire greater control, often as a means of protecting existing economic activities. All parties have their own reasons for pursuing consensus. For legal and political reasons, however, many watershed efforts focus heavily on land use and deal inadequately with instream flows. These efforts often are not well connected to state programs for water resource planning and streamflow protection. Strengthening these connections may be difficult in the current political climate that is increasingly hostile to government natural resource management. Progress in such basins as the Methow, Dungeness-Quilcene, Lemhi, Henry's Fork and Upper Clark Fork offers some hope, but it is too early to declare victory in any of these places. In considering the future relationships among ecosystem and watershed management, cooperative approaches to natural resource issues, and instream flow restoration and protection, a few cautionary notes are in order.

First, water law and water rights will impede progress. The prior appropriation doctrine, which is the core of Northwest water law, does not recognize the concept of watershed health. This doctrine allowed rivers and streams to be drained dry long before any state got around to protecting minimum flows. State water law and water rights will not likely be flexible enough to allow large-scale streamflow restoration, at least in the near future. And virtually every watershed management, instream flow, or basin planning effort is explicitly conditioned on protection of existing water rights.

Second, politics may change everything. The federal resource agenciest focus on ecosystems and watersheds may not last much beyond 1996. The current congressional majority clearly wants these agencies to focus primarily on commodity production. The same conservative political winds are blowing through Boise, Helena, Olympia, and Salem. Thus, federal and state agencies may place less emphasis on far-ranging resource protection strategies. On the other hand, obtaining consensus may become an even higher priority for federal and state agencies, in which case they are likely to rely all the more heavily on cooperative local approaches. Consensus may become harder to reach, however, as resource user groups believe they can get better results through the political process.

Third, what works in one basin may fall in another. The results of cooperative resource management efforts depend on many factors, including history, resources, personalities, circumstances, and timing. Success in a particular river basin often cannot be replicated because it is so closely tied to a particular place, time, and group of people. Cooperative approaches have helped ease longstanding conflicts in basins such as the Henry's Fork and Clark Fork. But they have had mixed results, at best, in other contentious basins such as Oregon's Umatilla and Nevada's Carson-Truckee. The optimism generated by progress in some basins is tempered by setbacks in others. Fourth, planning is important, but implementation is crucial. Translating paper gains into watershed improvements will prove difficult in every case. The potential problems—practical, legal, political, and fiscal, to name a few—are diverse, numerous, and often concealed. If these problems cannot be resolved, frustration with the plan and the process will set in, and even the most heartwarming cooperative efforts may break down. The Dungeness-Quilcene Plan recognizes these pitfalls at its conclusion:

This plan now needs implementation. That implementation must be integrated with Federal, Tribal and State and local watershed protection programs.... The immense effort put into this plan by an participants must not be wasted; the time is ripe to move forward in a coordinated effort to better protect and manage our water resources on the eastern Olympic Peninsula.

Fifth, these approaches are only now being tested. The majority of the Northwest's watershed planning and management efforts began only within approximately the past three years. Some of these newly sprouted programs, such as those in the Henry's Fork, Methow and Dungeness-Quilcene Basins, show promise but they have not yet had time to develop. Similarly, while various strategies exist to restore instream flows—some based on voluntary transactions, others on water law reform and enforcement—these approaches also have not been tried on a large scale. Some of these projects and good ideas might work, but it is too soon to tell. A facilitator of the Henry's Fork Watershed Council said, "I still consider this a big experiment. I hate to have people call us a model, because that implies we've figured it out."

In the Northwest, we are still figuring out how to resolve many difficult natural resource issues. But given our crisis over salmon, steelhead, and other fish, we clearly need new approaches. Experiments in watershed planning and management are certainly worth trying. If these experiments are to succeed, they must address streamflows effectively. Healthy watersheds, like rivers and fish, need water instream.

3

Water Quality Management Systems

Full efficiency could be achieved by water quality control measures at the individual waste outfall or water supply intake was useful for discussion of the maldistribution of costs resulting from technological external diseconomies, and probably descriptive of a significant number of real cases as well. But engineering-economic research his shown that there are many instances where economies can be realised by collective measures such as joint waste treatment, lowflow augmentation for quality improvement, stream reaeration, groundwater recharge for quality improvement, effluent diversion or redistribution to make better use of natural assimilative capacity, the specialised use of streams or stretches of streams, and combinations of these. Where facilities involve scale economies that cannot be efficiently realised at individual outfalls or intakes, it may be appropriate to have a regional authority with powers more extensive than those assumed in earlier.

At one extreme would be a river basin or sub-basin agency empowered to plan, design, construct, operate, and finance virtually all water quality control measures; at the other would be an agency that simply sees to it through effluent charges or standards-that downstream costs are reasonably well reflected in the upstream decisions of managerially independent units. In this part we emphasize the opportunities which a regional authority might have to improve the efficiency of water quality management by implementing various measures, such as those mentioned above, that cannot be efficiently instituted at individual points of waste generation and discharge. However, we also recognise the importance of the technologic and economic linkages between water quality improvement as an output from water resources systems and other outputs from such systems, such as hydropower, navigation, and water-based recreation. For example, operation of hydro facilities for "peaking" can reduce water quality significantly by causing extreme low flows during "offpeak" periods. Water released from the deeper parts of storage reservoirs which have "stratified" may be devoid of dissolved oxygen and have an effect on water quality downstream from the reservoir similar to the discharge of an organic waste.

Figure : *Clean-out and disposal for your Water Quality Management System*

The installation of low head dams for slack-water navigation may result in substantial water quality deterioration. The dams reduce the velocity of flow and hold waste loads longer in a given reach of stream; at the same time they reduce the reaeration capacity in the given reach. Conversely, the operation of a storage reservoir to augment flows during summer periods to improve water quality may have adverse effects on water-based recreation on the reservoir, because of the resulting drawdown. These interrelationships suggest the desirability of a regional water resources management agency with responsibility for all aspects of water resources development and use. We return to this point where we discuss more specifically the question of appropriate institutional arrangements for water quality management.

Within the context of both water quality management and overall water resources management, a major problem is to integrate in an optimal way the decisions of fiscally independent units with those made directly by a regional agency. An agency concerned with water resources does not possess the scope, even in principle, to decide all courses of action bearing upon the way water resources are used. For example, if such an agency decided that a firm should be located at a particular site because the costs of handling its wastes would be low at that site, it might force the firm to incur transportation or other costs greater than the saving in waste management costs. Owing to such possibilities and even probabilities, decisions on location, production, and the character of production processes should *never* be based exclusively on considerations of water costs including waste disposal. Nevertheless, it is important to develop means whereby these considerations can be brought to bear on the water decisions of firms and of public agencies and on the decisions of agencies with broader responsibilities for land use planning. It is suggested in this section that systems of charges have special merit for this purpose.

This chapter focuses on the economic criteria for planning a regional water quality management system that includes collective facilities as well as those that can be implemented efficiently at individual points of waste discharge or water intake and use. The chapter also includes some discussion of system operation because planning and operation are interdependent. The specific components of an optimal system are often quite sensitive to the operating procedures adopted.

A regional agency has a variety of potentially efficient collective measures to choose from: increasing streamflows during low-flow periods via releases from reservoir or groundwater storage; treatment of intake water and of municipal and industrial wastes in regional collective treatment plants in compactly developed areas; artificial or induced oxygenation of watercourses (including reservoirs and lakes); diversion of effluents via pipeline or other means; specialised stream use; artificial recharge of wastewaters to groundwater aquifers for quality improvement; and the construction of shallow oxidation reservoirs in the stream itself. These methods could supplement measures to control the generation and disposal of wastes at individual points. A regional agency seeking to minimise the overall costs associated with waste disposal should plan and implement a system which equates the relevant incremental costs associated with water quality management in all directions. These costs include the costs of all manner of waste reduction and treatment and water supply treatment, the costs of flow regulation (including all opportunity costs that arise from the value of stored water for alternative uses), as well as the costs of conforming waste discharge to streamflow, changing industrial processes and outputs, and changing industrial location, and, finally, any damages which result from residual water quality deterioration. This statement that the agency should equate the "relevant" marginal costs does not mean that the marginal costs of waste reduction should be equal at all points of waste discharge. Indeed, in an optimal system, as explained in earlier, the costs of treatment or other waste reduction measures will tend to be higher at upstream points because upstream activities are reflected in downstream quality and costs. What the rule does mean is that it must not be possible to make marginal "trade-offs" and thereby reduce damage and waste reduction costs.

Ideally, the results of all relevant water resources systems and operating procedures would have to be considered and a solution derived which simultaneously indicated the optimum combination of systems elements and operating procedures. In industrial areas the solution would entail a system of charges and/or other measures such as effluent standards and zoning designed to secure the optimum amounts and locations of waste discharges. An "efficient" solution would enable the maximum net benefit to be obtained from the available water resources, and, as the latter implies, the overall cost associated with the disposal of an optimum amount of wastes would be minimised.

In economic terms "cost" means the value of foregone opportunities. In regard to water quality management such opportunities may be associated with resources used, for example, to build and operate treatment plants or reservoirs which could be used productively in other applications, or they may consist of opportunities to use water for purposes other than waste carriage or assimilation. The development in this chapter is simpler to visualise if the only output of the system is water quality control, so that all costs associated with waste disposal consist either of resources devoted to control facilities (reservoir regulation of streamflows, treatment of wastes, process changes, effluent diversion facilities, etc.) or of damages associated with water quality deterioration. The principles are exactly the same, however, if some of the costs consist of foregone benefits from other water-related outputs such as peak power.

The Basin-Wide Firm Again

The basin-wide firm is used here to illuminate economic principles. It is not put forth as a policy recommendation, for it would, in reality, have grave disadvantages: It cannot take account of uses that are not reflected in market prices; and, as a monopoly, it would present problems of regulation.

To illustrate some of the principles involved in taking account of the technical interdependencies in a river basin when economies of scale in water quality management measures are available, let us assume that a single firm (1) conducts all water-using industrial enterprises, all water and waste treatment facilities (no privately owned septic tanks, water softners, etc.), and all water transportation and related facilities; (2) operates all hydroelectric facilities, owns all land and structures in the flood plain, and is the sole provider of flow regulation; (3) controls all recreational uses of the watercourse; and (4) operates in competitive markets or in ones where public regulatory authorities set prices equal to marginal costs at levels of output that just clear the market.

In order to maximise its profits the firm would select the combination of water quality control measures (water supply treatment, waste generation and treatment, flow augmentation, wastewater releases coordinated with streamflow, etc.) and downstream damages that would minimise the overall costs associated with its waste disposal at its most profitable level of activity. This would be accomplished by equating the relevant incremental costs associated with waste disposal for all alternatives. As previously indicated, these costs would include any valuable outputs that are sacrificed to improve the efficiency of the waste disposal system.

If collective treatment costs are less than treatment at individual discharge points, the firm would transport effluent to collective treatment plants until the marginal costs of transportation and marginal treatment saved were equal. Since transporting wastes long distances and building scattered treatment plants are usually expensive procedures, the firm would weigh these and other diseconomies of scattered development against the economies (less congestion) to arrive at the optimum compactness of economic activities.

Compact economic development might be influenced by certain physical characteristics of the stream. Since the rate of reaeration through the air-water interface is directly proportional to the oxygen deficit, the organic waste degradation capacity is a decreasing function of the dissolved oxygen quality standards. This, plus the possibility of stream treatment to take advantage of economies of scale, emphasizes the possibilities of stream specialisation (i.e., using some stream or stretches of stream more heavily for waste disposal and maintaining higher quality in others) as a means of reducing the costs associated with waste disposal. Again the firm would trade off such possibilities against alternatives such as additional flow augmentation, higher levels of treatment, and other means of waste reduction. In deciding upon the location of a new industrial plant, the firm would consider not only transportation costs and other factors affecting assembly, production, and distribution costs but also the increment of total waste reduction and damage costs of alternative locations. The location producing lowest costs would be the one at which the marginal costs of the alternatives bearing upon overall costs are equalised.

While the point is expanded later, it is worth noting here that the firm "internalizes" all these costs. A water resources agency in a basin cannot internalize all of them, and *consequently it must attempt to have the opportunity costs of water resources use, including waste disposal, reflected in the decisions of entities that are simultaneously considering other resources costs.* This can be done by standards or, preferably, by charges.

The firm's general objective is to maximise profits, which implies that the costs associated with waste disposal will be minimised at the profit maximising level of output. If, for example, the firm could lower its costs by doing a little less waste treatment and permitting a little more damage, or by doing a little more water treatment and a little less waste reduction, or by a little more augmentation of low flows and a little less temporary storage of wastes, or a bit more process adjustment and a little less direct reaeration of the stream, etc., its overall profit position could not be at a maximum. As part of its general profit maximising activities, the firm would integrate its waste control activities with other aspects of its water-related operations. One major linkage is through flow regulation.

Thus, in computing the costs of alternative water quality control devices, the firm would need to consider complementary and competitive relationships among different water uses and alteration of the stream's flow regimen. Accordingly flow-augmentation costs would have to be determined in light of the fact that this alternative is often complementary with navigation, for example, but at least partly competitive with irrigation, flood damage reduction, power, and recreation. Any net benefits associated with other uses foregone in using flow augmentation to improve water quality in the optimum system are counted among the costs associated with water quality improvement. Moreover, the firm would consider the full marginal costs of producing particular products including the costs imposed on other activities by waste disposal. If the market adequately registered the population's evaluation of all goods and services sold by the hypothetical firm, its solution to the waste disposal problem would be "efficient," as that term is understood in economic welfare theory.

Many extreme assumptions have been made in developing this example; hence the basin-wide firm is not put forward as a realistic policy alternative. Still, the concept has provided a useful vehicle for illustrating applicable systems and decision criteria. The following points concerning public policy merit particular emphasis. First, public policies which explicitly recognise water quality management as a problem of optimising an interdependent system in a region will find many alternative control measures to consider. Accordingly, a least-cost system is more likely to result from the implementation of such policies than from the adoption of narrowly circumscribed con ventional approaches. The case studies in the next show the savings that can result from a more comprehensive and flexible approach.

Second, implementing the regional approach requires the planning, design, and integrated operation of many interdependent elements in the system. The illustrative firm "internalised" all these elements and, under our assumptions, found their optimum planning and operation in its interest as a matter of seeking maximum profit.

A water management agency should not attempt to internalize more elements than needed for its objective. In one instance, an agency may plan, design, and operate only collective measures such as reservoirs and reaeration devices and use direct controls or incentives (charges) to integrate these with the rest of the relevant economic system. In other instances a regional agency may find that it serves efficiency to go much further and it may plan, design, and operate virtually all water-related facilities in a region including water and wastewater treatment plants. Even in this case it would still be necessary to influence industrial process design and plant location. This could be done by direct regulation or, again we think preferably, by an appropriate system of charges.

In the following section we develop these thoughts further by using a somewhat simplified example which nevertheless permits us to address some matters of planning incentives and financing in a system which includes collective measures.

Regional System with Flow Regulation

We assume a system where the economically relevant alternatives are low-flow augmentation and waste reduction activities at individual points of waste discharge. The regional agency will design and operate a reservoir for flow regulation and use a system of effluent charges to control waste discharges at outfalls. If the reservoir is multipurpose we assume that the agency takes appropriate account of complementary and competitive uses of the water in determining the marginal cost of flow regulation for water quality management. For example, if navigation benefits occur in a fully complementary fashion with water quality benefits, the two must be summed and compared with the marginal costs of flow regulation.

Determination of the optimum amount of storage can be illustrated by a set of curves closely analogous to those used to illustrate optimum levels of waste reduction. Consider again a hypothetical streamflow-cost curve. Costs in this case are the total costs at various levels of flow of optimal combinations of downstream alternatives-for example, waste treatment, water supply treatment, and value of residual physical damages. We assume that these combinations are induced by an optimal system of effluent charges in accordance with the analysis presented.

The effect of flow regulation is determined as follows. Starting with a natural (unregulated) flow frequency curve, the effect of different levels of regulation on the flow regime is estimated. In this case, however, the downstream costs associated with a given flow remain the same, but flow regulation changes the frequency of occurrence of different flows. As before, the integrals of the curves are the expected values of the probability of costs. From a succession of such calculations a total downstream expected value of costs-avoided function can be constructed analogous, but in this case the damage reduction corresponds to increases in minimum flow rather than reduction in residual waste material.

A total cost function for the storage capacity necessary to produce the increases in minimum flow is also shown. Minimum expected total cost (or maximum expected net benefit if avoided downstream costs are termed benefits) is achieved at an increase in flows corresponding to point *X*. The slopes of the total storage cost and downstream cost functions are equal to their respective incremental functions. Accordingly at point *X* the optimality condition (i.e., equality of incremental costs) prevails.

The same solution can readily be stated in terms of functions which are more familiar from previous discussion. In general the optimality criterion is that the *expected value* of the incremental costs of all relevant alternatives should be equalised at the margin. If effluent charges are used to control individual waste discharges, the optimal condition requires that such charges reflect the incremental offsite costs imposed by each waste discharge. Charges will of course be lower when the assimilative capacity is increased by flow augmentation and/or by stream reaeration. In a correctly planned system the lowered charges reflect a gain in efficiency.

It may be worthwhile restating why point *X* represents an optimum combination of upstream and downstream measures. At this point a small further increase in reduction activity at the outfall would cost more than the expected value of downstream water treatment costs and residual damages avoided by downstream users. The argument is symmetrical for a slight decrease in reduction activity at the outfall.

When costs associated with waste disposal are minimised, the incremental costs associated with waste disposal are equated at each outfall. This means that the marginal costs associated with waste disposal are equalised in all directions. However, it does not mean that the marginal cost of reduction will be the same at each outfall along a stream. This would only be the case if downstream costs per unit of waste were equal for different outfalls, as they might be when two outfalls are across the stream from each other; then strict marginal cost equality would be necessary for cost minimisation.

Effluent Charges and Collective Measures

Fortunately, devising an appropriate effluent charges policy for an agency which incorporates economically desirable collective measures in its system requires no new principles whatsoever. The ideal charge will still be equal to the marginal external cost imposed. The residual external marginal cost imposed by waste discharge, when control at relevant points of waste discharge is induced to an optimal level by effluent charges, is the measure of the marginal benefit of the collective quality improvement facility say a flow-regulating reservoir.

Let us assume (unrealistically) that the agency creates the entire assimilative capacity of the river by producing a flow of water. Assume further that this flow can be produced at a constant marginal cost. Finally, assume that assimilative capacity can be increased efficiently in small increments. Under these assumptions, the agency, if acting efficiently, would increase assimilative capacity in such a way that the marginal cost of the last increment in capacity to control the effect of a waste would be equal to the effluent charge on that waste, which in turn is equal to the marginal cost of waste control at outfalls our "equal marginal cost" principle again. In this instance the proceeds from the effluent charge reflecting the marginal external cost would just cover the total cost of producing the assimilative capacity.

Given the same situation, except that we now assume that the marginal cost of further increments of capacity is rising, the relationship between the yield from charges and the total cost of the investment will change. The yield from effluent charges will now exceed the amount necessary to pay the total cost of the investment when optimum capacity is attained. Conversely, if the marginal cost of expanded assimilative capacity declines over the relevant range, the proceeds from the charge will be insufficient to cover the cost of the capacity expansion.

This will be true even though the capacity expansion reduces the overall costs associated with the regional waste disposal system. If the regional agency decides not to augment assimilative capacity but to reduce waste loads by measures such as collective treatment, it should still follow the equal marginal cost rule and charge waste disposers the marginal reduction cost (which equals the marginal external cost) of handling the particular kind of waste involved.

The assumption that the regional authority created all of the assimilative capacity of the stream makes for a theoretically neat outcome, but it is unrealistic because a stream always has some natural waste assimilative capacity. This circumstance in no way affects the use of the marginal cost rule as the appropriate criterion for the agency's investment decisions, but it does affect the financial results. When the agency levies a charge reflecting marginal external costs in the presence of natural assimilative capacity, part of the payment made is a "pure rent" on the scarcity value of the natural capacity. As we have indicated earlier, this is analogous to a charge levied on the use of public lands for grazing, for example.

Accordingly, when natural assimilative capacity exists, the proceeds from the charge will be more than sufficient to pay for an economically justified augmentation of assimilative capacity produced under constant cost conditions. If an agency implemented only economically justifiable facilities and levied charges equal to the marginal external cost, it might take rather sharply decreasing costs in the provision of collective facilities for the agency to incur a deficit if there is a large amount of naturally occurring assimilative capacity initially available.

The above discussion has assumed that collective facilities can be expanded more or less continuously. Since some of them, such as reservoirs, involve substantial indivisibilities, this is not very realistic for small systems. But the principles are the same even when durable and lumpy investments are made. Of course it will be necessary to make long-run projections of population and industrial production and their related waste discharges with their attendant uncertainties. Other measures such as mechanical reaeration can be expanded more or less continuously. Groundwater recharge operations represent an intermediate situation. Moreover, as a system becomes larger and more diverse, even absolutely large additions become small relative to the total system and can be implemented on the basis of comparatively short-run considerations.

Unmeasured Values as Constraints on the Objective

If there were no problems in the manner in which the market registers water quality values, the policies described above would produce a solution quite consistent with the general rationale of a market system. But the market provides no value, or no satisfactory value, for some aspects of water quality, and this creates a special problem for public bodies because they must take account of all aspects. Considerable progress has been made in finding ways to assess the worth of such matters as general aesthetic effects, public health, and recreation, but dealing with these unmeasured values is still far from being a routine matter.

Two ways of handling unmeasured values come to mind. One way would be to label them "intangible" and disregard them in planning and designing water quality management systems. Then, when the proposed systems are presented for consideration and authorisation by representatives of the public, side information could be provided on aesthetic effects, public health, and other matters considered relevant to arriving at a decision in the public interest.

The second approach is to include hypotheses about such values in the process of system planning by expressing them in physical terms and treating them as constraints upon the cost-minimisation objective. For example, if social choice should dictate that the oxygen level in a stream must be high enough to support fish life at all times, the system must be planned to minimise the real cost associated with waste disposal subject to the constraint, or "minimum D.O. standard." This may require a different combination of units with different operating procedures than a system planned without the standard or constraint. If the constraint is effective and not automatically achieved when measurable costs (including damage costs) are minimised, it will increase the cost of the system, the extra cost representing the limitation which the constraint places upon the objective. To be consistent with efforts to achieve maximum welfare, the planning process should consider constraints "provisional" and view them as matters for research and study in order to discover how well they represent the preferences of society. One way of studying them from this point of view is to test their cost sensitivity. Varying a constraint by small amounts, redetermining the optimum system, and relating the change in costs to the associated physical changes in specific stretches of stream will provide information that will permit considered choices to be made by political representatives.

One useful way of stating the results of experiments with the constraints that are not valued directly by, or imputable from, the market is in terms of what they must "at least be worth." For example, a social judgment may be made that dissolved oxygen is to be maintained beyond the point indicated by the cost-minimising solution in order to preserve or enhance a recreational fishery. A comparison of the system with and without the constraint will not establish precisely what preserving the fishing pleasure is worth, but it will indicate the *least* value that must be attached to it if the higher level of control procedures is to be worthwhile. Or the additional cost of a higher level of certainty of achieving the desired dissolved oxygen level can be estimated. If constraints are imposed representing goals not directly commensurable with the values stated in monetary terms in the objective function, marginal conditions analogous to those indicated earlier must still hold if the cost-minimisation objective is to be fulfilled. The optimum system is not attained until a situation is reached in which it is impossible to make incremental "tradeoffs" between alternatives that will lower costs without violating the constraint. Once a combination of system elements has been decided upon, it will be necessary to utilise charges, effluent standards, or some other regulatory device to achieve optimum waste reduction at individual outfalls. The charge, if a charge is used, must be just high enough to achieve the standard, and it must not be possible to expand the large-scale measures at an incremental cost less than the incremental cost which can be avoided at the individual outfalls. If several wastes are involved, the expected value of all incremental costs avoided by an expansion in the large-scale measures must be added together and compared with the cost of expanding the measure.

We may generalise by saying that once a constraint is established by a duly constituted, well-informed political decision-making process, it may be viewed as representing a damage function which is perfectly inelastic at the specified quality level and degree of certainty. Large amounts of data and extensive knowledge of physical and economic relationships are needed to implement the procedure outlined above. Rapid and flexible computational techniques for estimating the physical characteristics and patterns of water quality and for carrying out the actual minimisation (maximisation) procedure are also essential. Some of these techniques, as well as the specific character of the information required to make reasonably dependable analyses of complex river basin systems, including cost sensitivity analysis of constraints, are illustrated by the studies described in the next.

Here we simply note that once a system is implemented the charges themselves are important sources of information about investment decisions in collective facilities. They indicate the marginal cost which waste disposers are incurring in order to meet the standard. This information can be used by the basin agency to determine whether a collective measure, either by capacity expansion or improved operation, can maintain the standard at a lesser incremental cost. Thus once the system is established the agency can obtain information on costs incurred by waste disposers without having detailed knowledge of their cost functions.

Institutional Restraints

Where there is no regional agency with authority to plan and operate a comprehensive system of quality management, the type of cost-sensitivity analysis outlined above can be used to illuminate the desirability of institutional change. In most basins in the United States the only collective measures are industrial and municipal waste-treatment facilities and reservoirs operated for low-flow augmentation by federal or state agencies. By comparing a least-cost system restricted to these measures with one that incorporates all relevant alternatives, a price tag can be put on the restriction. Or, to put it another way, the comparison will provide an estimate of the economic gain that society can derive from an institutional mechanism with authority to implement a wider range of alternatives.

Some "Short-Run" Aspects of System Operation and Cost Assessment

At this point it is useful to introduce once more the distinction between "long-run" or "planning" costs and "short-run" or "operating" costs. There are two reasons for making this distinction in the present context: First, a system should be operated on the basis of its "current" opportunity costs, especially when certain elements such as dams must be introduced in "chunks." Second, hydrologic or meteorologic fluctuation may make it desirable, in terms of cost minimisation, to operate the water quality control facilities in a variable manner. This indicates that, aside from any informational and administrative costs, effluent charges should be varied so as to maintain equality in the short-run marginal costs of alternatives.

So far the discussion has been primarily in terms of planning the regional system, and it was assumed that choices were not limited by past decisions. The view taken was sufficiently long-run, for example, to permit the building of dams *instead* of treatment plants. For planning, these are valid and appropriate concepts.

The point has been made several times that if the objective is to minimise the value of resources used internally and externally in connection with water quality management, the relevant cost for choosing a particular alternative is the value of the opportunities that must be foregone if it is chosen. This concept of "opportunity" costs has been used throughout, and methods of producing a better reflection of the value of foregone opportunities to society in decisions with respect to waste disposal were the subject.

The *concept* of opportunity costs does not change when a particular set of facilities has been decided upon and installed, but the *character* of the opportunities foregone does. When a dam is in the planning stage, all the labour, cement, steel, etc., needed for its construction can still be used in other activities, but once the labour, cement, etc., have been embodied in a dam they can no longer be used in alternative activities, and their price can no longer be considered an opportunity cost. The only costs that are still opportunity costs and thus relevant for further decisions are those that arise when a dam is operated for a particular purpose. If the dam is multipurpose, the opportunity costs may be largely "internal," representing the net value of the other uses foregone when its capacity is used for a particular purpose, although some operating costs will be involved, too.

In planning water quality management systems, therefore, the definition of costs should include all capital costs of new facilities. The system for water quality improvement should be expanded to the point where the cost of reducing an *additional unit* of any given waste or combination of wastes if the reduction activity involves joint products raises total reduction costs as much as it diminishes water quality damages. If there is a constraint, the system should be expanded until the constraint is met and the marginal total costs of all alternative measures for achieving the constraint are equalised. However, after a water quality management system is established, only the currently variable costs and opportunity costs internal to a multipurpose system are relevant. For example, assume that a decision is to be made about whether to retain a waste treatment plant with excess capacity or replace it with a smaller plant. Comparison of total costs, including capital costs of both plants, would almost certainly lead to the wrong conclusion. If the original plant has no alternative use, the relevant costs include only the operating costs, the land, and portions of the existing investment for which there are alternatives uses. The value of such uses must, of course, be brought into the comparison because they represent genuine opportunity costs. The test is always whether there are currently feasible, valuable alternatives. From the point of view of allocating resources to their most productive use, only inputs which have alternative uses can be said to involve costs.

Since storage reservoirs ordinarily cannot be economically constructed in such a way as to add small increments to system capacity, a system adjusting to changing waste load conditions will have periods of excess capacity. Under these circumstances the short-run opportunity costs of using the reservoir for flow augmentation for water quality improvement must be compared with the opportunity costs of alternative uses of the reservoir capacity. Appropriate operating procedures of the system would be based upon relevant marginal costs. Reliance upon flow augmentation would be heavier in the early period after the dam was built, and effluent charges or standards would be accordingly lower in those years. This means that care must be taken to provide waste disposers with information on longer-run prospects so that investment decisions will not be based on the presumption that charges will continue to be low. This chapter has focused on the integrated planning of various types of waste load reduction, flow regulation and in-stream measures; the incorporation of "unmeasured values"; and some "short-run" economic problems which stem from environmental variability. Attention was directed to the important role that can be played by effluent charges-either alone or in conjunction with other measures for regulating waste discharges in articulating the decisions of individual waste dischargers efficiently into a regional system for water quality management. In addition, it was pointed out that the assessment of charges is important in financing the collective features of such a system.

It was indicated that, ideally, system planning would consider all alternative uses of water, including its use for waste disposal, the effects of the various uses on water quality, the losses imposed on other uses by quality deterioration, and the value of water-derivative uses. All feasible system plans and operating procedures would be considered, and a solution would be derived which would indicate the optimum combination of system elements and operating procedures in light of the objectives and constraints relevant to the system. The regional system view, taking cognizance of the efficiency of collective measures, did not require the introduction of any new economic principles. But there are practical reasons, including the inherent complexity of the problem and the primitive state or absence of much of the required data, why an optimum solution cannot be fully attained. Furthermore, some inevitable arbitrariness in the specification of constraints, the absence of fully satisfactory means of handling the risk inherent in natural phenomena, and the uncertainty involved in forecasting future economic variables, militate against the achievement of fully optimum decisions. But with ingenious simplification, it is possible to develop workable procedures and mount investigations that will produce satisfactory, if not totally precise, results.

Nevertheless, an approach of the general type outlined in this chapter will improve decisions on water quality management. It emphasizes the importance of identifying interdependencies and viewing areas and economic functions tied together by "spillovers" in a "system" context; requires an explicit and unambiguous statement of objectives and constraints; stresses the identification and quantification of alternatives and their systematic testing in light of criteria of merit developed in terms of the objective or objectives of the system; and endeavours to cause the opportunity costs of water resources use to be reflected in decisions not controlled directly by the water resources management agency.

These features of the analysis represent a way of thinking about the problem of water quality management; they do not imply a specific procedure for solving it. When the situation is a complex one involving numerous alternatives and constraints, optimisation procedures applied to formal mathematical models, often with the aid of computers, can be very helpful.

4

Challenges and Realities of Water Management of Megacities

Megacities—characterized as metropolitan areas of 10 million or more inhabitants—have become an important focus in terms of water provisions, sanitation services and the related impact of urban development on natural resources. While basic needs of residents of smaller cities are very similar, the emphasis placed on megacities lies in the fact that these massive urban conglomerates have grown to almost unmanageable dimensions. This, in turn, has made water provision and sanitation services to the entire population a bleak, if not unachievable, task. In most megacities of the developing world, clean water is both scarce and expensive to produce. Large sectors of the population in such densely populated urban areas do not have access to potable water, and even larger sectors of the population do not have access to sanitation. Non-revenue water—water that is produced and enters the distribution system but that is never billed to consumers because it is lost due to leakages or illegal connections—is up to 30 to 40 percent. Infrastructure is either scarce, becoming complex or deteriorating. In addition, water supplies are largely underpriced and necessary investments are estimated to reach billions of dollars. However, as complex as these matters are, the real challenges in terms of water provision point in another direction. The main problem in urban concentrations, which is further heightened in megacities, is the lack of appropriate management, adequate institutions and sustainable planning to address these challenges beyond short-term approaches.

To provide an overview, this study first analyzes the nature of megacities on a global scale, addressing the delicate relationship between economic development and urban resource management in the face of environmental sustainability. This topic then unfolds through an analysis of the Metropolitan Area of Mexico City (ZMCM), a megacity accounting for approximately 20 to 22 million people and recognized for its diminishing supply of natural resources, among them, freshwater. Emphasis is placed on declining groundwater supply, complexities with infrastructure and concerns regarding sanitation. While the central focus of this argument is the ZMCM water supply; a comparative analysis points out the similarities and differences among megacities across the globe, such as Dhaka, Sao Paulo and Bangkok. Ultimately; this chapter attempts to demonstrate that the problem of water access in megacities has less to do with water scarcity and more to do with water mismanagement.

Megacities as a Growing Phenomenon

As urban growth continues to increase worldwide at escalating rates, megacities remain a growing phenomenon. According to the Organisation for Economic Cooperation and Development (OECD), as globalization progresses, urban regions—in particular, cities with populations exceeding 10 million people—are playing increasingly important roles in the global economy in terms of competitiveness, thus becoming major centres for both local and national policies. Yet, most megacities have focused on economic growth with almost no consideration for environmental sustainability.

As a result, these extensive metropolitan areas have created alarming levels of water and air pollution, solid waste and deterioration of their own natural resources. These problems have also reached surrounding regions affected by urban spraw. The most serious environmental problems are associated with the disorganized expansion of legal and illegal settlements, over-exploitation and degradation of surface and groundwater sources, air pollution, solid waste and the indiscriminate destruction of natural resources. This has created what seems to be an endless vicious cycle in which the deterioration of the environment has resulted in social erosion and vice versa. Mexico City and its surrounding areas have not been an exception.

While Mexico City has achieved high economic growth, its development is still deeply polarized and asymmetric. Economic and social development has flourished in some parts of the megacity resulting in prosperity, but inequalities have become increasingly acute in other parts of the region. In order to maintain and improve the living standards of a growing population, major investments are needed to protect basic resources. Failing to address environmental degradation—such as shrinking freshwater sources—is counterproductive to economic development in the long-run. This holds true because while most megacities focus on increasing foreign competition and employment at the cost of their natural environment, foreign investors tend to be most interested in megacities that not only have the most appropriate institutions, infrastructure and human resources, but also liveable environments and natural resources that will sustain economic activities.

Although all megacities face similar challenges in terms of access to natural resources—like clean water and ensuring adequate control of air, land and water pollution—these urban conglomerations differ in terms of growth rates, types of population, infrastructural needs and environmental conditions. In the developing world, metropolitan areas tend to have high population growth, consistent needs for infrastructural development and increasing deterioration of their environment. In contrast, in the developed world, megacities tend to have slower urban growth, infrastructures that already need to be replaced and more regard for the protection of their environment. In both developing and developed megacities, migration of formal and informal settlers plays an important role in terms of demand and provision of services. In Mexico City and its surrounding areas, illegal settlements are a critical problem. In 2003, there were 804 irregular settlements in Mexico City with approximately 60,000 families. In Mexico State—the most populated of the country's thirty-two states—illegal settlers are creating mushrooming cities that lack basic services. Much of this population is established not only in valleys and hills, but also in areas prone to flooding, which can result in flood hazards that can destroy entire communities. Legal and illegal settlements in flood-prone areas and their social and economic related costs have been heavily felt in several other megacities in addition to ZMCM, such as Sao Paulo and Dhaka. In most cases, it is a problem of which control and solutions go beyond the water sector and which is vested in land-use planning institutions.

Due to their differences in terms of population, location, climate, infrastructure and environmental concerns, each megacity needs to plan and implement its own tailored solutions at its own pace according to local economic, social, environmental and cultural conditions. Yet, a common denominator uniting megacities—mainly in the developing world—is that rates of urbanization have surpassed the capacities of both local and national institutions to provide the population with the necessary services. According to the UN Population Division, the number of megacities has increased from sixteen to eighteen from 2000 to 2005 alone, and this number is predicted to reach twenty-seven by 2025. It is noteworthy that the majority of megacities at the global level are located in the developing world.

In spite of urban growth, the percentage of urban populations living in megacities worldwide has remained about 8 percent from 2000 to 2005. On the contrary, it is in urban conglomerates of less than 500,000 people where the percentage of urban population at the global level is still the highest (52 percent) by cities with 1 to 5 million inhabitants (22 percent). This means that, despite the rate of increased urbanization throughout the last decades, the highest percentages of the world population do not live in megacities; they are more likely to live in small or medium-sized cities. While the growth of small and medium sized-cities is projected to escalate and, with it, the challenges of providing adequate water services, it is the megacities that have turned to almost unmanageable dimensions and require a reexamination of water management.

Furthermore, megacities are not restricted to their geographical boundaries but extend to the regions surrounding them. The escalating demands of megacities in terms of clean water, among other basic necessities, have resulted in negative impacts on the surrounding states and regions, which have been forced to supply megacities with additional commodities. Since megacities continue to increase in size, demand requires the implementation of development models based on social equity and environmental considerations, not only at the local level, but also at the regional level.

Despite these bleak realities, alternatives exist to improve the governance of megacities in order to provide citizens with better access to clean water and other services, as well as improved environmental conditions. OECD, for example, suggests that governance-related issues should be addressed at the metropolitan area level in order to maximize the potential of the urban region. Some of the actions mentioned to improve metropolitan governance include encouraging local governments to work in partnership with business communities, introducing greater transparency and accountability in decisionmaking, developing better frameworks for long-term strategic planning, and improving public access to information and public services, all of which could lead to positive results. The limitations that OECD identifies for achieving better governance at the metropolitan level, however, include a fragmentation of the administrative jurisdictions, a strain on the financial and fiscal ability of local authorities as well as the lack of decisionmaking processes that are both transparent and accountable. There is also a lack of ability to coordinate and effectively communicate between cities that form part of metropolitan areas.

Complexities of Water Resource Management in Megacities

From the late 1990s to the present, water management trends have included, at least in theory, the implementation of concepts such as sustainable development, river basin management and integrated water resources management approaches. An emphasis has been placed on multi-sector approaches, as well as higher considerations for social and environmental issues, with increasing importance given to governance-related issues.

In the water sector, attempting integrated approaches has made water management goals even more complicated to implement than before, particularly in the context of megacities due to their sheer size. With the added expectations on implementation of water governance issues, even more complex processes have been introduced—such as transparency and accountability, or multilevel participation and multilevel decisionmaking—even when the most basic and pressing needs in terms of water supply and sanitation have not yet been solved.

Megacities require large investments to cover the needs of all sectors of the population in terms of water supply, drainage and sanitation. However, in addition to the enormous investments, a main constraint for the efficient provision of water supply is that governments have multiple, fragmented institutions with responsibilities that are often not clearly defined or overlap with each other. This makes any long-term decisionmaking almost impossible, leaving strategic planning in the utopic realm, resulting in escalating degradation of the urban environment.

Although considered basic needs, water supply, drainage and sanitation are not always an investment priority in megacities. This is because lack of water, both in terms of quantity and quality, does not affect the entire population to the same degree. In general, wealthier populations have access to tap water in their houses, as well as functioning drainage and sewage systems, while less affluent sectors of the population depend on standpipes or purchase their water from vendors. Many poor regions are either connected to a public sewer or septic tanks, or have to dispose their wastewaters in surrounding rivers, lakes or even on their own property. Lack of tap water and drainage in megacities, therefore, is a problem faced by less affluent sectors and is therefore their problem alone. Sanitation is an increasingly serious issue which only a select few communities have to face directly but which affects the health of the entire region.

As previously stated, water management also faces the complexity of flood-prone zones. For example, the number of humans exposed to floods has tripled during the last two decades to almost 2 billion at present. The main reason behind this problem is the concentration of hundreds of millions of people in mushrooming cities in both deltas and floodplains. On the other hand, megacities sprouting in arid and semi-arid regions face water scarcity on a daily basis, not only because of mismanagement but also because of an actual water scarcity problem to begin with. Regarding water for irrigation, megacities devour enormous amounts of provisions which have to be transported from the countryside, often at great distances.

Megacities and the Escalating Case of Water-related Challenges

The debate on water resource management in the international arena most often focuses on integrated water resources management, river basin management, good governance and public participation. In spite of this rhetoric, priority in megacities is the provision of clean water even at increasing social, economic and environmental costs.

Investments for water supply and sanitation in the developing world, including megacities, has escalated considerably. In the case of Mexico, water supply and sanitation investments were estimated to be $1.7 billion per year in 2002. In 2005, China invested a total of $9.7 billion on flood control, infrastructure, soil, water and ecological projects and hydropower development. At this point, many scholars and water activists have pointed out the failures of achieving the development goals at the international level; cities provide water that is not always potable, and collection of sewage has been conveniently considered as sanitation even when it is not treated.

Varis et al. summarize the situation and future prospects of seven rapidly expanding megacities in various parts of the developing world: Jakarta, Johannesburg, Silo Paulo, Mexico City, Riyadh, Istanbul and Singapore. In their analysis, the authors conclude that megacities are dramatic cases of urbanization and water-related challenges. A main constraint is that even though the provision of water for various sectors is quite important, water is not the only aspect of infrastructure development needed in megacities. Transportation, energy and housing should also be developed.

The water footprint of a megacity goes far beyond its city limits. Megacities import food from the international food trade, the bulk of which has been produced using water irrigation practices.

Megacities also import massive amounts of other resources such as energy, metals and fiber products, which can interrupt hydrologic systems, both qualitatively and quantitatively. In order to be sustainable, megacities need flourishing economies, strong social and environmental policies, proper governance systems and adequate mechanisms for public participation. All of these aspects are intertwined with the various roles that water plays in megacity development, and the water sector is often an important contributor.

Finally, economic growth, as well as social and environmental factors, should be considered simultaneously Competitive megacities are better able to attract resources to invest in infrastructure, thereby providing quality services for the population and, in turn, better economic and social opportunities for large sectors of their population. The challenges to achieving this are immense. However, initiating the necessary processes and frameworks for financing, managing, maintaining and providing efficient services, as well as finding long-term solutions that combine social, economic and environmental view points, is necessary. Otherwise, freshwater access and sanitation become unmanageable.

Water Governance in the Mexico City Metropolitan Area

Applying this overview of megacities to a case study, this section offers examples of the complexity of clean water services faced by the ZMCM. It is important to stress that, more often than not, inefficient, costly and unreliable water supply services in Mexico City are the result of inappropriate management practices, not merely the result of an actual scarcity in water supply. To reiterate the size of this megacity, the ZMCM has an approximate population of 20 to 22 million people. It generates approximately 35 percent of the national GDP in Mexico, which makes it strategically quite important from both an economic and political standpoint for the country. High rates of migration, in addition to rapid urban development, have made Mexico City not only one of the largest cities in the world, but also a city with one of the highest population growth at the global level. The phenomenal growth rate witnessed between 1950 and 2005—from 2.9 million people to more than 20 to 22 million at present—has made it almost impossible for institutions to respond to the rapidly increasing needs of the population in terms of water supply, drainage and sanitation.

The rapid growth of the ZMCM has been the result of policies that have favoured social services, long-term major investments in all sectors (energy, water, transportation, telecommunications, etc.) as well as heavy subsidies and industrial production. Even though government planning strategies have been focusing on decentralization efforts for decades, tax subsidies, as well as numerous social, political, educational and cultural factors, have made the city more attractive than other regions within Mexico.

The multiple levels of public sector institutions that are involved in the solution of water-related problems make decisionmaking a very complex process not only in the ZMCM, but in all megacities. In Dhaka, a city with approximately 12 million inhabitants, organizations responsible for water supply include government institutions at the federal, city and municipality levels, the private sector and even donors. The World Bank, for example, was the main donor of the Fourth Dhaka Water Supply Project and insisted that changes in governance and management be implemented at both the Dhaka Water Supply and Sewerage Authority (DWASA) before the credit was approved. In the case of the Silo Paulo Metropolitan Area, decisionmaking in water-related issues for their 18 million people involves federal, state and municipal levels of the city of Sao Paulo, as well as the additional thirty-nine municipalities that form part of the megacity.

The ZMCM includes two different political entities and their several levels of government. Decisionmaking depends on decisions based in the sixteen boroughs of Mexico City, thirty-seven municipalities in the Mexico State—each municipality consists of an independently elected representative—and the private sector companies which are also involved in service provision. Another important issue is that from 1997 onward, the citizens of Mexico City were able to elect their own government. Prior to this, the mayor of Mexico City was appointed directly by the president, who, from 1929 until 2000, belonged to the same political party The government of Mexico City, therefore, has often been considered an extension of the federal government, with criteria and priorities mirroring that of the federal government. This means that, for better or for worse, there was agreement on the policies that would be followed in the metropolitan area. Currently, the different political parties represented in the governments (federal, state, boroughs in the case of Mexico City and municipalities in Mexico State) have diverse interests, priorities and criteria. This diversity, while unquestionably positive in terms of democracy, has resulted in a more serious lack of coordination and agreement in terms of economic, social and environmental policies. The reality is that political criteria prevail even when this is not always positive for the megacity.

Groundwater Supply in Mexico City and Beyond

The added complexity of the ZMCM for its water management, in addition to its large population and fragmented government, is that it is located in a naturally closed basin, making it vulnerable to floods. In addition, its own rivers have become wastewater canals. Furthermore, increasing subsidence has resulted in extensive damages to the water supply and sewerage infrastructure, as well as degradation of the quality of the groundwater on which its own survival depends.

The water supply in ZMCM will provide for 20 to 22 million people; the economic sector depends primarily on local groundwater sources (70 percent) and on inter-basin transfers from outside basins. Infrastructure for distribution of water within the city has become very large and complex. It is comprised of sixteen dams having a total storage capacity of more than 2,800 cubic kilometres, a primary network with more than 1,000 kilometres and a secondary network with more than 12,000 kilometres.

In order to supply the necessary water for the needs of the social and economic sectors of the megacity, the rates of extraction from the aquifer have been much higher than the recharge rate for decades. At present, approximately 45 to 54 cubic meters per second is abstracted per year, but the natural recharge rate is only about half of this figure. This has resulted in very serious overexploitation, with the groundwater table decreasing by around one meter each year. It is estimated that approximately 670 legal wells exist for groundwater extraction.

Likewise, the megacity of Dhaka, Bangladesh, depends heavily on groundwater. The water supply system for the city that was constructed more than one hundred years ago was based on surface water. However, the cheaper and easier option was the availability of groundwater, which resulted in a supply system based on groundwater. At present, with surface waters heavily polluted, institutions face the challenge of providing services to a population increasing at approximately 6 percent per year, while simultaneously trying to reduce dependence on groundwater. This is a system that is clearly unsustainable.

Available data indicate that the groundwater table is falling at an alarming rate of more than 2 meters per year. Groundwater is being mined in significant amounts. Consequently, the discharge of the tube wells are being reduced at a much faster rate resulting in higher cost of pumping and drying up of the bore holes. The life of bore holes has been reduced from twenty years in the late 1960s and early 1970s to less then five years now. Investment in new deep wells is quickly becoming economically unattractive.

Deficit between supply and demand is expected to widen again from 2004 but at a much faster rate due to the inability of the DWASA to secure necessary funding for the construction of two more surface water treatment plants. At the same time, construction of new and replacement old deep tube wells will not produce designed discharge as the groundwater reservoir of both the primary and secondary aquifers have nearly been exhausted. Though a new initiative of exploiting deep aquifers (more than 300 meters deep) have yielded some promising results in a certain area of the city, any large scale development of the deep aquifer will have to be carefully evaluated for its technical, economic and environmental viability.

According to official figures the water extracted from the aquifer to supply water to the population of Mexico City from 1990 to 2007 has reached almost 17,000 cubic meters per second. This figure is significantly higher that the groundwater recharge rate, illustrating how this trend in groundwater extraction is unsustainable.

The problems related to the extraction of groundwater in Mexico City are not new. Several issues relating to the city's survival were identified decades ago. Since soil is composed of clay, the more water that is abstracted from the aquifer, the more susceptible the soil is to compaction. Subsidence of several parts of the city has resulted in extensive and costly damages to both surface and below ground infrastructure, including water supply and drainage.

The ZMCM is not the only megacity which suffers from severe subsidence. Bangkok is also affected by subsidence in certain areas of the city. In the case of Bangkok, subsidence rates are estimated to be 1.5 to 2.2 centimeters per year as of 2001, a much lower figure compared to that of one decade ago which was 5 to 10 centimeters per year. In comparison, the central area of the ZMCM has subsided approximately 10 meters during the 1900s, varying from 10 centimeters per year in the central area to 20 to 25 centimeters in the area of the International Airport of Mexico City.

Both megacities have established mitigation measures in the attempt to reduce the abstraction rates of their aquifers. These have included policies, economic disincentives and, in the case of ZMCM, protection of conservation or rural areas for natural recharge purposes. However, conservation areas are under increasing threat because of steady growth in both urban and rural areas. Regarding rural growth, 35 of the 44 rural settlements that still exist in Mexico City and which include 400,000 people, are also located in conservation areas.

Access to Tap Water in Megacities

In terms of access to water, in the developed megacities, such as in London and New York, all residents normally have access to tap water in their houses. In the developing world, a common problem in most megacities is that millions of their inhabitants do not have access to clean water in their houses or nearby. These populations depend on either standpipes, vendors or provision of water from tankers provided by the public sector. In these cases whether or not residents are consuming good quality water is uncertain.

In ZMCM, as in the Silo Paulo Metropolitan Area, approximately 90 percent of the population has access to water, either through a water connection directly to the house or from common faucets in the neighbourhood. In the case of ZMCM, most of the water sources are located to its west, north and south. Thus, the water supply is irregular and unreliable for the population living in the eastern part of the city, the subsection most affected by water shortages. An alarming figure is that in Jakarta, only approximately 41 percent of the population had access to tap water in 1996, a figure which increased to 56 percent in 2003 despite the changes in the institutional arrangements and the involvement of the private sector.

Non-revenue waters are one of the main challenges facing megacities both in developed and developing countries. The Office of Water Services (OFWAT)—a water regulatory agency of England and Wales—provided Thames Water (a private sector company) a target to reduce water losses to 805 million liters per day for 2006-2007.

In Mexico City and its surrounding areas, non-revenue water is approximately 30 percent. In the case of Tokyo, on the other hand, non-revenue water accounts for only 9 percent. Lack of investment tends to be the most frequently mentioned reason for high, non-revenue water. However, lack of managerial skills and manpower, as well as inefficiency and poor planning, are also considered major causes for the low performance of utilities providing the water supply.

Unreliability of water quantity results in extra expenditures for certain sectors of the society in ZMCM. In fact, it is estimated that the population without access to tap water pays 6 to 25 percent of their daily salaries in purchasing 200-liter containers of water. Nevertheless, mistrust in the quality of Mexico City's tap water also has negative impacts in terms of extra expenditures for all sectors of the society irrespective of their economic status. Drinking water for most of the people in the ZMCM, even when they have access to tap water, comes mostly from 20 to 30 liter containers of drinking water that are sold commercially. A main problem, however, is that companies that bottle water in the ZMCM are not monitored and thus water quality of the containers is not always in accordance with the specific norms and regulations.

The perception that the quality of tap water is not suitable for drinking purposes has an enormous impact on the economy of the country. Mexico is the second largest consumer per capita of bottled water in the world, with consumption increasing from 11.6 billion liters in 1999 to 17.7 billion liters in 2004. At present, per capita bottled water consumption in Mexico is nearly twice that of the United States, even though its per capita GDP is about one sixth that of the United States.

The Third World Centre for Water Management has carried out analyses on water quality in Mexico. According to this study, in 2005 alone, the expenses due to waterborne diseases in the country were estimated to be $260 million, with the highest mortality in population between zero and four years of age and above sixtyfive years of age. Nevertheless, despite the economic importance of the impacts that water-borne diseases have had on the country for decades, these findings have not been documented thoroughly and addressed properly by health and water-related institutions. In summary, the problems related to the drinking water supply in megacities include multi-dimensional, multi-sectoral and multi-regional issues, which depend on the interests and capabilities of more than one institution and which require multi-stakeholder coordination.

Wastewater Collection

In addition to water supply, the other main challenge for megacities in terms of water is collection of wastewaters and their proper treatment and disposal. It is said that 2.6 billion people lacked improved sanitation facilities in 2002, which is a number too big to be ignored by governments and by the international community. Not even the United Nations considered sanitation important enough to be included in the Millennium Development Goals. It was not until the World Summit on Sustainable Development organized in Johannesburg in 2002 that sanitation goals were considered as a part of a global effort to improve the quality of life of the populations as well as quality of the environment.

In the ZMCM, about 50 cubic meters per second of wastewaters are disposed without treatment. Almost all of this water is used for irrigation purposes, resulting in very significant health and environment-related problems and concerns. In 1996, the Inter-American Development Bank approved a $1,035 million project for the Mexico Valley Sanitation Project. Unfortunately, this much-needed project did not proceed, mainly due to economic and political reasons. In 2004, water institutions from Mexico City, Mexico State and institutions at the federal level were working jointly with the Inter-American Development Bank (IDB) and the Japan Bank for International Cooperation (JBIC) to develop the terms and references related to the construction of four wastewater treatment plants. The total budget for this project was approximately $1 billion, of which IDB would contribute $365 million for the collectors system, the JBIC would provide $410 million for the wastewater treatment plants and the federal government, Mexico City and Mexico State governments would provide $260 million.

This enormous sanitation project was abandoned for years by the federal government, Mexico City and Mexico State authorities, being reconsidered only recently. Seven treatment plants, instead of the seven mentioned initially, are now expected to be constructed with a total capacity of 40 cubic meters per second. The approximate investment will be $2,345 million. Only time will tell if the project will be completed.

Energy Consumption and its Relation to Water

Concerning energy, megacities require massive quantities of power. Megacity residents consume about five to ten times the energy of the national average. In addition, all large-scale electricity generation requires an enormous amount of water, either as hydropower or for cooling purposes, as do bioenergy sources, making water an important prerequisite to satisfy the energy, requirements of megacities.

For the ZMCM, the daily consumption of fuel in 2004 was estimated to be 306,000 barrels equivalent of gasoline or petrol, or 48.6 million liters per day. In 1990, the estimated figures were 37 million liters per day, that is, there was an increase of 31 percent in fourteen years. Drinking water to Mexico City has to be pumped to a height of more than 1,000 meters. Equally, because of the increasing soil subsidence, wastewater also has to be pumped up to discharge from the city. The above has made water supply and sewerage very energy-intensive and expensive. In 2000, for example, energy used in the ZMCM on pumping clean water, treating water for drinking purposes and collecting wastewater and rainwater was 2.436 Peta Joules. As it can be observed, the necessary energy needed to pump clean water has cost the country almost $600 million per year between 2000 and 2005.

Strategies for Drinking Water and Sanitation: Lagging Behind the Needs

Strategies to increase water supply and sewerage services have focused mainly on infrastructure development throughout the years. The different strategies (e.g., the 1995 Master Plans for Drinking Water and for Sewerage for Mexico City) have outlined the necessary infrastructure to improve supply, storage and transportation of drinking water in Mexico City, as well as the storage, transportation and disposal of wastewater and stormwater.

For several years now, the program on drinking water for the ZMCM has considered constructing new infrastructure to transfer more water to the metropolitan area from even more distant and expensive sources. These alternatives have considerable economic, social and environmental costs.

In the case of the transfer of water from Cutzamala River, the construction of its fourth stage would have an initial investment of $502 million to increase the volume of water only by 5 cubic meters per second, from 19 at present to 24 cubic meters per second. If the water transfer from the Amacuzac River was implemented, in addition to the cost of the infrastructure, the annual electric power consumption for this system would be about 5 percent of the annual national electricity production, representing 16.5 million barrels of oil per year. In terms of investment, each cubic meter of water from the Cutzalama River is estimated to require an investment of $23 million. This figure would increase by a factor of four if the source of water were the Amacuzac River. This is a very clear example of the costly impacts that the disorganised growth of the megacity has imposed in the surrounding regions, creating acute economic, social and environmental problems.

Considering the above, it will certainly be more acceptable from the economical, social and environmental view points to implement demand management practices like reduction of unaccounted losses, water pricing and other water conservation practices, before embarking upon extremely expensive new water development projects, with high social and environmental costs.

It is estimated that the large volumes that are lost due to leakages in the distribution systems in the ZMCM—representing between 30 and 40 percent of the total volume of water that it is distributed—are second only to human consumption. It would thus be preferable to focus first on repairing the distribution networks before embarking on such expensive projects, especially if water would still not reach the consumers due to the losses.

As Iracheta mentions, water in the metropolitan area is both scarce and expensive. In spite of this fact, water is not reused or treated after use but discharged raw, with the associated environmental and health-related problems, to another basin. Unfortunately, even then, long-term and rational planning, as well as coordinated policies for the development and management of the metropolitan area, including water resources, is still not the priority for the federal government or the government of Mexico City and Mexico State.

In 2003, the government of the Mexico State filed a lawsuit against the federal government and the government of Mexico City. This occurred as a result of the lack of response from the federal government and city government to the request of the government of the Mexico State to renegotiate the agreements to transfer water to Mexico City from one of its sources, the Lerma River. The lawsuit demanded the payment of $2.5 billion as compensation for the use of water from the Alto Lerma from 1970 up to that moment. It also demanded that the federal government become responsible for the operation of the infrastructure, which is operated by the government of Mexico City since the construction of the projects even though they are located in Mexico State.

Finally, it demanded that the federal government implement actions to recharge the aquifers. Even though Mexico City pays the federal government for the rights for the bulk water it receives from the Mexico State, the lawsuit from the Mexico State asked for compensation for the environmental damages that have been caused due to the overexploitation of the Lerma River. The decision of the Supreme Court was expected to set precedents for similar cases in the future. However, in October 2005, the newly elected governor of the Mexico State publicly declared that he would withdraw the lawsuit, since he preferred to work with the federal and Mexico City governments to find an amicable solution.

Solutions on the Horizon: When Efficiency Overcomes Size

While the size of the geographical area does matter, so too does the planning and effectiveness of decisionmakers. There are two urban conglomerates in the developing and developed world, respectively, which have lately improved the provision of water supply and sanitation to their populations to levels higher than any city in the United States or Western Europe, irrespective of their size in terms of population. These are Phnom Penh, Cambodia and the country of Singapore.

In the case of Phnom Penh, in 1993, the non-revenue water was 70 percent, the tariffs (as is typical in all developing country cities) were lower than the cost of water produced and the water utility faced serious economic and human resources constraints. Their billing system was quite inaccurate and the actual volume of water billed represented approximately 28 percent of the production. From those actually billed, the collection ratio was only about 50 percent. Finally, the total annual income was only about 0.6 billion Rids, against an operating cost of 1.6 billion Riels.

By 2006, the Phnom Penh water utility developed an operational billing and collection system through which the customer database increased from about 20,000 in 1993 to almost 160,000 in 2006, and the collection ratio increased to 100 percent. The tariffs have been increased to more realistic values, and non-revenue water is now less than 10 percent (lower than London and similar to Tokyo). The difference in Phnom Penh was achieved as a result of leadership, political support, managerial skills and human resources development. The water supply in Phnom Penh provides clean water now to 100 percent of the population, a striking difference from only ten years ago.

Singapore is an exceptional example of what planning and management, as well as long-term vision, can achieve in a period of time as short as forty years. Singapore, being a water-scarce country due to a limited amount of land where rainfall can be stored, has developed—and implemented—innovative solutions that have included both demand and supply management strategies. Demand management strategies have included water transfer from the state of Johor in Malaysia, management of their reservoirs, desalination and development of very high technology to treat used water and produce very high quality water, known as NEWater. Supply management strategies have included conservation practices, tariffs to promote water conservation and large public education campaigns. Overall planning and management strategies developed in Singapore have not been matched anywhere in the cities of the developed world, irrespective of their size.

It has been said that solutions in Singapore can be implemented because of the small size of the city-state. Nevertheless, supply and demand management strategies are, or should be, common practice everywhere, especially in megacities where the needs to properly manage the water resources available is more pressing due to the size of the population that has to be served. For water policies to be useful for the cities and their population, these management strategies have to be implemented. In only a few cases is there the will, the skills and the vision to implement them. Singapore is one of these few cases.

The supply of clean water and sanitation services in megacities should be viewed as an integral component of rapidly expanding urban areas. Clean water and sanitation services should not be considered in isolation, but instead in relation to issues such as their geography, climate, population growth, urbanization, migration, economic development and social expectations. While most of the cities in the developed world are growing slowly, cities in developing countries are growing in a disorganized and unplanned manner, resulting in heightened pressures to provide water supply and sanitation services to their ever-growing populations.

Provisions of services do not only include operational alternatives, such as the development of massive infrastructure, which is a necessary, but not always ideal, alternative. They also involve the development of a very complex series of issues that have to be adapted to each case, which include management aspects, governance concerns and decisionmaking adaptations—mostly shared between several actors—to suit the local conditions, develop a basket of economic instruments as alternative measures and encourage human resources skills to accelerate socioeconomic development.

An issue that is normally disregarded, but is of high importance, is that economic activities and naturally available water resources—rainfall, springs, lakes or rivers—do not always go hand in hand. In addition, most megacities have either exploited or polluted their surface and groundwater resources, making it necessary to transport water—many times hundreds of kilometres—at increasingly higher economic, social, environmental and political costs in order to cover their own needs.

Finally, megacities are faced with challenges that require immediate action. One of the most important challenges is governance. Governance-related issues are not just the domain of the government any more. The increasing participation of the private sector and societal actors—who are willing, and also demanding, to participate in solving problems that affect them—has made decisionmaking very complex. These complexities increase exponentially within the context of a megacity. Nonetheless, improving water governance in megacities should be of the utmost concern.

Water Management in the Upper Little Colorado Watershed

Perhaps nowhere in the United States has climate and water resource interaction been more closely observed than in the U.S. Southwest, and with good reason. Overall, semiarid to arid conditions predominate. These conditions, together with wide variability in temperature and precipitation over time and space, challenge even the most earnest human efforts to sustain livelihoods and communities. Scientific studies and popular literature alike reflect recurrent episodes of overexploitation of natural resources due to inadequate recognition or acceptance of environmental limits such as water availability.

Accretion of local knowledge and experience provides a mechanism for anticipating and adapting to the vicissitudes of variability and change. However, factors such as reliance on faulty heuristics (Nicholls 1999), occurrence of events falling outside the range of local memory, policies that restrict decision options, and narrow focus on immediate economic returns can render such knowledge ineffective at best and destructive at worst. Science-based knowledge and predictive capabilities provide avenues for improving decision processes in the face of interacting environmental and societal stresses. To be accepted and used appropriately, such knowledge needs to be franked in ways that mesh well with local values and practices. Environmental histories that create a contextual account of embedded experience, values, and practices provide insights useful to this task.

Recent extended dry conditions in the U.S. Southwest have prompted efforts by researchers and forecasters to develop climate information products useful at local and regional scales for addressing drought impacts. This chapter provides an environmental history of climate and water resource management for the Upper Little Colorado watershed in northeastern Arizona, with a focus on the hundred years between the mid-nineteenth and mid-twentieth centuries. The study, funded through the NOAA-funded Climate Assessment for the Southwest (CLIMAS) project, provides insight into the experiences, values, knowledge, and practices that have produced conditions existing today. This type of contextual information, if integrated into development of plans, decision processes, and decision tools, holds promise for improving citizens' acceptance of and support for policies aimed at ensuring the sustainability of the watershed's closely coupled human and natural systems.

Based on established historiographic methods, this study aims specifically to uncover how communities and individuals in the watershed have developed and managed water resources in the context of climate variability, especially drought. We present a brief discussion of the precolonial land-and resource-use patterns, and then examine patterns, practices, and events of the past two hundred years. Though we briefly discuss contemporary conditions and trends, we focus on the period between roughly 1850 and 1950. This time period represents early settlement and resource exploitation practices, the more extensive Mormon colonization period of the late nineteenth century, and significant changes that occurred in the ensuing decades. Our analysis considers local experience as well as how local entities have negotiated water shortages in the evolving context of state and national laws and institutions.

Mormon settlement and agrarian practices are especially important to consider in examining conditions in the watershed today. These first Anglo-American settlements profoundly affected subsequent patterns of population distribution and resource use. The period of population expansion following initial Mormon settlement taxed the water resources available within the basin, and prompted development of local knowledge as well as adoption of successive technologies to address water scarcity problems. Cultural values that emphasized agrarian lifestyles' even while acknowledging that the region was far better suited for animal husbandry and grazing than for farming, strongly influenced the choice of strategies for coping with water resource challenges. Rather than adapting to grazing lifestyles typical of arid areas, residents built dams and introduced groundwater pumping as strategies for supporting crop cultivation. This approach still prevails today.

Post-World War II introduction of mining operations and power-generation facilities into the watershed, combined with rapid growth in tourism and recreation, compounded the stresses on water resources in the basin. By the late twentieth century, water demand throughout the basin combined with recurrent drought led to rising concerns about climate vulnerability. While recognition of local authority over water development and use persists, momentum has grown for introducing institutional mechanisms at the state and federal levels to increase adaptive capacity. Provision of scientific information about climate, hydrology, and resource management and promotion of structured planning processes are among the most prominent contemporary efforts to improve decision making at the watershed and local levels in the U.S. Southwest. In the process, the region's environmental history will be influenced as well.

The Little Colorado River Basin: A Case Study

The evolution of economic, social, and cultural structures and dynamics in the Upper Little Colorado watershed reflects processes unfolding throughout the western United States. Occupied by indigenous peoples when the Spanish explorers first arrived in the early 1600s, southwestern landscapes and livelihoods were significantly transformed by Spanish introduction of livestock grazing and development of towns beginning in the mid-1600s, followed by expansion of farming activities with the arrival of Anglo-American settlers beginning in the mid-1800s and increasing after the end of the U.S. Civil War in 1865. Mormons wrought some of the most influential landscape transformations during this era through their extensive efforts to establish farms across the region, wherever growing seasons were long enough and sufficient water was available.

The arrival of railroad lines in the late 1800s produced another round of profound socioeconomic changes across the larger region, prompting rapid expansion of livestock herds, introduction of mining operations, and decimation of forests to provide energy for early mining and industrial needs. Cities such as Denver and Phoenix grew rapidly, especially at the end of World War II when families across the nation began moving westward in search of better lives. The growth of these urban areas, together with increased interest across a broad range of the U.S. population in natural-area tourism and recreation, introduced a new dynamic to rural areas: temporary—and often large—influxes of people during the prime months of summer (for hiking, fishing, etc.) and winter (for skiing and other winter sports), as well as growth in the number of cabins and homes owned by residents of the urban centres and, increasingly, by retirees.

Innovations in Internet technologies over the past decade and a half have allowed people to work in remote areas, further swelling the number of permanent residents in some towns. These changes, in turn, have fueled growth of rural towns and the number and variety of services offered in these towns. Meanwhile, across the entire time span from the arrival of the Spaniards to the present day, the native peoples of the western United States have sought to sustain their cultures and livelihoods, often under conditions of considerable adversity. They lost large portions of their previous domains and found themselves largely marginalized from the dynamics of the dominant society, yet their lands, practices, and experiences continue to stand as testaments to pre-European ways of knowing and being. These landscapes also testify to the profound disparities in social and economic well-being that exist among many tribal members. The Upper Little Colorado watershed reflects these conditions and dynamics.

Social Constructions and Biophysical Realities

As historian Elliott West maintains, there is always a dissonance between how humans view the world and how they actually fit within it. Humans have the ability to imagine changes in the world even as they confront it: they visualize new connections and relationships that are not yet present (West 1998). The American West is one of the places where imagination has crucially influenced changes to the natural world, producing a series of imagined landscapes. Here, broad swathes of arid and semiarid landscapes have been transformed into irrigated agricultural landscapes and urban settlements. The Little Colorado River Basin, whose modern settlement patterns grew out of Mormon visions of Zion and its related agrarian landscapes, is a microcosm of social constructions of the arid/semiarid West. The landscapes we see today, indeed, are the products of what historian Mark Fiege calls "vivid metaphor and compelling myths".

The Little Colorado River watershed provides a useful focal point for investigating the environmental history of relationships among climate, natural systems, and human systems: it is characterized by significant climatic and related environmental fluctuations, and it gathers into an identifiable geographical space a coherent narrative of human-environment relationships over time. The watershed, 26,964 square miles in size and ranging in elevation from some 4,200 feet to more than 7,700 feet above sea level, is located in the northeastern corner of Arizona. The area is characterized by topographic complexity, including elevational changes that produce spatial variability in temperature and precipitation and related ecological variability. While the area's climate is characterized by high levels of variability over decadal and longer time spans, climate-division data reveal long-term temperature patterns of cold winters and mild summers and a bimodal precipitation pattern of rather wet winters and summers, with dry spring and relatively dry fall seasons. These data show an average annual temperature for the area of 55[degrees]F, with average lows dropping to 7[degrees]F in winter (in the highest elevations) and ascending to highs of around 80[degrees]F in the summertime (Sheppard et al. 2002). It is notable, however, that over the past century overall temperatures have risen approximately 2.5 degrees. Though highly variable over time, precipitation averages approximately thirty-six inches a year at higher elevations, and about eight inches a year at the lowest elevations.

Scientific advances in the past several decades have improved understanding and predictive capability with regard to one of the major winter climatic influences on much of the Southwest: the El Nino-Southern Oscillation (ENSO). During El Nino years, the region tends to get wetter than average winters, while during La Nina years, winters are likely to be dry. Current forecasting capabilities allow skilful prediction of possible ENSO conditions as early as the fall preceding the winter rainy season; climatologists update these forecasts monthly based on measurements such as trends in sea surface temperatures in the tropical Pacific Ocean. The climate of the summer months, on the other hand, remains difficult to predict very far in advance. Unlike the broad-scale precipitation associated with winter storm fronts, summer climate conditions are characterized by highly localized, often intense thunderstorms associated with the North American Monsoon. As noted previously, the spring season tends to be very dry, but may include significant rain events triggered by synoptic-scale conditions such as tropical cyclones.

Paleoclimatic research highlights longer-term trends in the region's climate, including multi-decadal periods of drought alternating with periods of near-average or above-average precipitation. Such shifts have profoundly affected natural and human systems in the Little Colorado River Basin since the arrival of the first humans in the area.

Most of the land use in the basin is rural. Indian lands encompass almost half the territory, and only about 23 percent of the land is in private ownership. Forests of blue spruce, Douglas fir, and ponderosa pine dominate the higher elevations, interspersed with grasslands. Drier, lower-elevation areas of the watershed are characterized by sparsely vegetated desert scrub landscapes.

The Little Colorado River, with its tributaries, is the only significant source of surface water in the region. There are, however, many natural seeps and springs. The river, which is generally narrow and shallow and is fed by some thirty smaller tributaries, extends 250 miles between the White Mountains and the Black Mesa Highlands, eventually emptying into the Colorado River. Snowpack from the White Mountains is the primary source of water to the stream; in addition, water from the Coconino Aquifer, also flowing south from the White Mountains, surfaces as a series of springs along the main stem.

The water resources and vegetation coverage of the area are both influenced by climatic conditions, which are marked by high inter-annual to decadal and longer-term variability, as well as by geographical variability across the watershed. Winter precipitation and related snowpack conditions are crucial to sustaining both surface water and groundwater resources in the basin. Rain during the summer season also contributes to surface water availability although, due to the fast-flowing runoff produced by strong storms, summer precipitation is not a major contributor to groundwater aquifer recharge.

Paleo-environmental History of the Watershed

Pre-Columbian Native American land-and water-use practices provide our earliest evidence of interactions between climate changes and corresponding cultural adaptations in the region. The Ancestral Pueblo, Hopi, and Navajo peoples all developed their own water management and shortage-mitigation strategies that, unlike drought adaptations in the nineteenth and twentieth centuries, hinged on mobility and community interconnectivity. Historians and archaeologists have pieced together a complex, evolving narrative of the earliest known habitations in northern Arizona, with most agreeing that humans entered the watershed between 30,000-11,000 BCE and 9500-9000 BCE (Gumerman and Gell-Mann 1994; Jones and Tagg 1994).

Archaeological evidence indicates that a mixed strategy of hunting, gathering, and horticulture prevailed by at least 600 BCE, and that in many areas, settlements were seasonal in nature. Responses to drought or flood events revolved around simply moving some distance away from the affected areas. With a shift toward greater agricultural complexity, populations grew, as did the use of available resources. The growth did not occur along a straight line, however. An archaeological study of the Silver Creek watershed revealed significant population increases in the region around AD 900-950 and AD 1000-1050, but found a decline in population between AD 1050 and 1100. At least some of the fluctuation in demographic numbers seems to derive from a change in settlement patterns from small to large: the study found there was a population decline in small sites and an increase in large sites during AD 1100-1150 (Mills, Herr, and Van Keuren 1999).

While scholarly disagreement persists regarding the temporal organization of increasing settlement complexity in the Southwest, most agree that by AD 1000 to 1150 human population growth and settlement densities had produced high levels of intercommunity interaction (Gumerman and Dean 1989). Relatedly, increased reliance on agriculture for subsistence increased local vulnerability to climatic variability, since the agrarian groups were less free to leave their crops and move to other areas than were hunter-gatherers. With reduced mobility, the Ancestral Pueblo peoples, for example, changed their practices to emphasize reliance on intercommunity interactions and mixed subsistence patterns (Gumerman and Dean 1989). Communities along the Little Colorado River appear to have adopted a local system of shifting agriculture. For example, groups inhabiting the Homol'ovi area north and east of present-day Winslow, Arizona, shifted their agricultural activities in response to climate-influenced environmental conditions. As local population expanded and individuals were better able to support larger-scale irrigated agriculture, floodplain agriculture came to predominate along the Little Colorado River. During low river flows, however, these farmers were still able to grow crops using runoff and seepage fields, on stabilized sand dunes to the east of the floodplain (Lange 1998).

At the Ancestral Pueblo settlements in nearby Chaco Canyon (located in present-day northwestern New Mexico) agricultural and timber resource demands from a growing population severely taxed the ability of the environment to sustain a large population centre.

A dry period between AD 1080 and 1100 probably caused many of the inhabitants to seriously consider the efficacy of unchecked demographic growth in the canyon (Lekson 2006). Agricultural choices would have been limited, for as population grew, ever more marginal land was brought under cultivation. With the advent of deep sustained drought in the twelfth and thirteenth centuries, the capacity of the land and water resources throughout the Colorado Plateau to support large, sedentary settlements diminished radically (Dean 1988).

Not much is known about activities in the region between the era of Ancestral Pueblo decline and the arrival of the Spaniards, though it is clear that the Spaniards changed the ecology of the region considerably through the introduction of livestock.

Research suggests that the Navajo people entered the area in the fifteenth and sixteenth centuries, perhaps reflecting a strategy of migration from the Rocky Mountains of Colorado and New Mexico to escape the drought that occurred in that area during the late 1500s (Bailey 1980). Navajo subsistence practices revolved around hunting (by men) and agriculture (by women). Though some evidence of granaries suggests the existence of surpluses, trading between settlements was likely to occur only when climatic conditions favoured sufficient surplus production (Brooks 2002). The introduction of Spanish livestock from Spanish settlements on the upper Rio Grande in the seventeenth century resulted in adoption by the Navajo of sheepherding; the demands of pastoralism, in turn, led to a dramatic expansion in the people's use of the land and its resources, as well as to social stratification and military aggressiveness (Bonte and Galaty 1991). At the same time, pastoralism also concentrated people and animals into the areas where water and forage were in greatest abundance (West 1998). As resources were depleted, Navajo residents migrated west and south to new lands, leading to pressures on the neighbouring Hopi and Apache populations (Mauss 2003). The drought of 1736-48 prompted yet another southwesterly migration (Brooks 2002).

Euro-American Arrival, Mormon Settlement, and New Environmental Stresses

Although Spanish expeditions entered the area in the late 1500s, the Upper Little Colorado watershed saw little non-native settlement until the arrival of the first Mormon settlers in 1876. These individuals brought with them an ethos of sedentary agriculture based on irrigation technologies and corresponding religious institutions that, among other things, determined settlement distribution, influenced land-and water-use practices, and provided a framework for dealing with drought.

The arrival of Mormon settlers was preceded by exploratory trips undertaken by Mormon Church members between the 1850s and the early 1870s. Church leaders, including Brigham Young himself, found the area to be acceptable for Mormon expansion and redemption of desert lands and, equally importantly, believed that the move would facilitate religious conversion of the area's Hopi and Navajo inhabitants. Young commanded specific groups, chosen for their devotion to the faith and their loyalty, to move from Utah to the Little Colorado watershed and establish self-sufficient communal orders (including communal ownership of water) that would be based on irrigated agriculture.

The settlers arrived at an auspicious time, preceded as it was by two five-year periods of wetter than average conditions: the period from 1833 to 1842, with a total of eighty-four inches of precipitation, was the second wettest decade on record in the region; the years from 1865 to 1869, with almost forty-four inches of total precipitation, constituted the third wettest five-year period. It was during the latter period that the Mormon Church sent the series of exploration parties to the watershed. Given precipitation conditions in the years before their arrival, it is perhaps not surprising that their reports depicted the area in favourable terms. It is interesting, however, that the settlers viewed the many Indian ruins in the area not as a warning about the dangers of overtaxing the environment with excessive human occupation, but as a sign that through their religious work the land could be resettled: "to our settling this new land, we had evidence that this land has been densly [sic] populated, this choice land will be blessed according to our faithfulness" (J. Smith 1881).

The reports encouraged Mormons to migrate to the area but also generated unrealistic expectations with regard to the region's agricultural capacity. Ironically, in their efforts to protect their religious beliefs and practices and avert competition over land and ideology (Journal of Discourses 1862), Mormons often sought out lands deemed undesirable by other people. In the Little Colorado watershed, as discussed in more detail later, realization of the full extent of the limits of the land and its resources emerged all too soon with the return of deep and long-lasting drought conditions in the 1890s. Throughout the colony, in local church and community meetings, the settlers talked of trials from God and the need to improve the land through communal endeavor and thus spread Mormonism to as many lands as possible (Leone 1974, 1979). Limiting the colony's success, however, was their philosophy, which focused not on how much rain fell or how much water was in the river, but on how hard they worked at improving the land.

The Mormons in this way were like many other western settlers who believed that a religious work ethic would transform arid land into agricultural paradise (H. Smith 1950). This logic dictated that through producing and sustaining a thriving agricultural community, they would achieve victory over natural forces. The settlements in Utah had proven that victory was possible: in 1879 Wilford Woodruff, future president of the Church, "spoke of the first settlement of Utah, that it was a desert, the country is now full, and we have not got room. Zion has to lengthen its cords, it matters not where we are.... It is our calling to build up the Kingdom of God" (Eastern Arizona Stake 1979-1886).

The topography of the region, which includes steep slopes and elevational changes that influence local ecological forms and processes, makes agriculture challenging at the best of times and under the best of circumstances. At the time of Mormon settlement, the best agricultural areas lay in the southern portion of the watershed; here water was most abundant and flooding was less intense. However, although many joined Jacob Hamblin in recognizing that the ecology of the watershed was, overall, best suited for grazing—sheep grazing was the primary land use in the area when the Mormons arrived—most Mormons acted on their desire to plant grains and thus improve the land to a "higher" form of land use. As one settler put it, "our people are an agricultural people and wish to beautify and adorn their homes and to improve and cultivate the land. I believe that the Lord has selected this new land for his Saints to gather to, as in thickly populated places wickedness is more prevalent.... In these new places there is a feeling of grandeur, a feeling that brings God nearer to us, and brings us nearer to Him" (Snow, quoted in J. Smith 1881). Creating this Kingdom of God required building dams and water-diversion projects, which would eventually render much of the river channel near the settlements devoid of riparian vegetation.

Upon their arrival in the basin, Mormon settlers identified two main areas for community development. One was in the middle of the river's length, in an area stretching from modern-day Winslow to the town of St. Johns, where a few Mexican settlers had already established homes near a post built by Solomon Barth to supply soldiers stationed at Fort Apache. These early settlers, who herded sheep and grew a little grain (Fish 1895[?]), were most likely immigrants from New Mexico who were attracted to the supply post. A small settlement also existed on Silver Creek, near present-day Snowflake, when Henry Stinson arrived and began building irrigation ditches there in the early 1870s.

The first Mormon settlers had little knowledge about the Little Colorado River or its flow regime, or about the ecological and climatic processes of the region. However, over the last quarter of the nineteenth century and first decades of the twentieth century, they built, in their church meetings, a body of local knowledge and series of strategies for responding to climatic stress. Their approach, which blended observations of local conditions with religious doctrine, fostered community cohesiveness at the same time that it sought to reduce vulnerability to environmental stresses. As settler Fredrick Christensen observed, "It [is] nescessary [sic] to learn by experience hence we should school ourselves and be prepared to return back to the presence of our Father. Let us stick to this work, for this is a preparation to go back and redeem the Centre Stake of Zion" (St. Joseph Stake Minutes, 1878).

The irrigation-based agricultural livelihood practices of the Mormons, especially during the early years before the introduction of water pumps and sources of energy to run them, dictated a settlement distribution pattern that was largely based on proximity to the river and the ability to take advantage of gravity to direct flows into the fields. Settlers' decisions about where and how to plant were also influenced by ecological conditions that had been produced by the previous series of wet years. Yet, with the onset of deep drought in the 1890s, much of the knowledge that had informed planting decisions had to be rethought. Lack of sufficient reservoir storage limited the range of options available to some farmers, while sporadic flood conditions forced others to deal with damage to farmlands and structures created by dam failures.

The spatial and temporal variability of stresses on the communities in the watershed reflected to a large extent, then as today, spatial and temporal variability in climatic and hydrologic conditions. This variability helps account for the fact that some of the Mormon settlements established with high expectations did not survive past the nineteenth century. The settlements of old Taylor, Obed City, and Brigham City were among the earliest communities established, having been built by the first group of migrants arriving in 1876, yet all were virtually abandoned before the beginning of the twentieth century. Obed, located amid wetlands, dissolved in the wake of a malaria outbreak in 1877 (Jensen n.d.). At old Taylor and Brigham City, repeated dam failures and resulting crop failures led to the communities' demise. Woodruff often borrowed food supplies and manpower from other settlements, due to its exposure to spring and late-summer floods: the settlement was located below the confluence of a seasonal tributary with the Little Colorado River. Between 1876 and 1890, the settlement experienced ten dam failures. Thirteen dam failures occurred near the settlement of St. Joseph over the same period. Archival information indicates that these downstream communities have always been more susceptible to flooding and water shortages than their upstream neighbours. People living in the downstream stretches of the river had to depend on surface water that was infamous for its unpredictability and its heavy silt load. These communities were eventually forced to resort to groundwater to overcome problems caused by irrigation activities on the main channel of the Little Colorado River. Dissatisfaction with communal institutions arose during the later decades of the nineteenth century, eventually resulting in their abandonment. The demise of the communal culture occurred in no small part because of the difficulties in sustaining lives and livelihoods in the region.

More broadly, though some settlements were successful and remain thriving communities today, the early Mormon experiences reveal the types of problems that arise when significant disparities exist between values and livelihood practices on one hand and inescapable environmental limits on the other. Many settlers, upon abandoning communities in the lower basin, moved upstream to establish new towns such as Snowflake and St. Johns. Here they found greater success, for they faced much less difficulty in developing surface water and groundwater resources. The settlement of Snowflake, for example, purchased from Henry Stinson in 1878, quickly became the most productive agricultural settlement in the region because of its relatively rich water resources. Silver Spring and Silver Creek were—and still are—the most consistent water sources in the basin. Rather than fluctuations in water availability, the primary challenge for the upper-basin settlements such as Springerville, Eager, Alpine, and Pleasant Valley was the short growing season due to higher elevations.

Yet even in these communities, concerns about water shortage persisted. In 1886, for example, the Joseph City Irrigation Company mandated a fee of $7 per acre for water used by its members. The revenues were to be used for building and maintaining water infrastructure (Mien City Minutes 1876-1886). These fees continued to rise as St. Joseph's water problems persisted: there was insufficient water when they needed it in the spring, and too much water when they did not need it in the late summer. By 1894, water fees had risen to $20 per acre, forcing individuals to plant more acreage (McLaws 1894).

With the arrival of the railroad in 1881, communities had access to more sophisticated irrigation and water-diversion technologies. At the same time, however, the railroad brought large-scale cattle grazing into the watershed. The relocation from Texas of the Aztec Land and Cattle Company (also known as the Hashknife Outfit), with herds of as many as 40,000 cattle, was particularly instrumental in driving environmental change. Mormon settlers strongly objected to the cattle operators' encroachment on and claiming of vast tracts of land, noting a lack of respect for the rights of others. The issue of encroachment was especially critical to the Mormons, for legal practices at that time persistently failed to officially recognize Mormon settlement claims.

A series of wet years in the late 1880s, coinciding with the arrival of the Hashknife outfit, prompted the group to ship in even more livestock and eventually to run 150,000 head of cattle and 120,000 head of sheep in the area. These animals shared the range with Mormon herds and with greatly increasing numbers of Navajo sheep, thus placing enormous stress on the area's rangelands. Conditions changed, however, due to deep drought beginning in the mid-1890s and extending to 1904. During this time period, the area received less than half the precipitation that had fallen during the previous wettest five-year periods. During the five-year period in the depth of the drought, 1900-1904, total precipitation amounted to a little less than twenty inches. The environmental and social impacts were huge. Mormon residents commented that this drought had "no parallel in recorded history" (LeRoy and Wilhelm 1982); water was scarce, and the grass disappeared. One settler observed, "It is stated that three fourths of all the stock along the river will die. The stock men feel about ruined and the outlook looks dark for all classes of business" (Fish 1895[?]). Widespread erosion and siltation of waterways were among the environmental effects of the interactions between drought and overgrazing. The comparison with conditions when the Mormons first arrived was stark:

> *When we came to Arizona in 1876, the hills and plains were covered with high grasses and the country was not cut up with ravines and gullies as it is now. This has been brought about by overstocking the ranges. On the Little Colorado we could cut hay for miles and miles in every direction.... Water followed the cattle trails and cut the country up. Later tens of thousands of cattle died because of drouth and lack of feed and disease." (quoted in Tellman, Yarde, and Wallace 1997).*

A study of arroyo-cutting in the Southwest in the late nineteenth and early twentieth centuries (correlating roughly with the years of Mormon settlement) found widespread erosion throughout the region. During this time, the Southwest was characterized by rapid and pronounced erosion as well as by the desiccation of river valleys. Social factors limited locally to the watershed, such as Mormon irrigation practices and cattle grazing by the Hashknife, constituted one but not the only driver of erosion (Cooke and Reeves, 1976). Indeed, erosion and arroyo cutting, especially on the Little Colorado, is best explained as the outcome of human activity combined with climatic variability and changes in vegetation cover. Maitland Bradfield (1971), in his study of Hopi agriculture, found that climatic variability and loss of vegetation contributed heavily to erosion in the Little Colorado River watershed between

1880 and 1900, 1971). He notes, however, that late-nineteenth-century arroyo cutting along the Little Colorado River and elsewhere can only be explained by combining the impacts of social factors with the impacts of climatic variability. Clearly, though, the 1890s drought contributed substantially to erosion processes as well as to corresponding water-resource problems throughout the Mormon settlements.

Twentieth-century Institutional and Technological Responses to Climate Variability

The impacts of the turn-of-the century drought on water resources prompted significant institutional change, including the first systematic effort within the basin to record the water rights of each holder. This effort foreshadowed modern water-rights institutions that organize rights to water chronologically, with the claimant having the longest history being accorded first priority, and so on. The information produced by the recording process prompted a change in water allocation practices in times of shortage: rather than relying on community decision making, water was henceforth allocated based on individual recorded rights. For instance, in Joseph City, on April 15, 1907, a committee of three was appointed to "record the dams and water according to their judgement" (Porter n.d.) The Joseph City Irrigation Company, previously responsible only for construction and maintenance of dams and irrigation ditches, was now charged with managing water allocations in response to drought-induced shortages.

These sorts of institutional changes also afforded opportunities for coping with variability within long-term droughts. For example, while 1904 stands as one of the driest years on record (with a mere 1.6 inches of precipitation), 1905 (at 13.7 inches) was the third wettest year since the late 1890s. Such wide swings in inter-annual variability made it virtually impossible to sustain consistent agricultural output by relying solely on surface flows. The first strategy adopted by the settlers for coping with variability in water availability was impoundment of water in reservoirs. The St. Johns Stake, for example, built seventeen dams between 1883 and 1921 on the main stem of the Little Colorado River and its tributaries (LeRoy and Wilhelm 1982). Dams and reservoirs posed their own challenges, however: violent summer storms could produce large volumes of water that led to breaches in the clams and consequent loss of life and property, while extended drought conditions diminished the amount of water entering the reservoirs.

Around the same period, the Joseph City Irrigation Company made the first serious attempt to systematically explore for and exploit groundwater resources (Porter n.d.). This effort notwithstanding, groundwater development in much of the basin remained minimal until the mid-1900s. As late as 1946, only 600 acre-feet of groundwater were being pumped in the Joseph City-

Holbrook area (Babcock and Snyder 1947). By contrast, groundwater pumping in Navajo County was much more intensive, producing an estimated 11,000 acre-feet around 1950 and more than 38,000 acre-feet by 1972. The growth in groundwater development coincided with three of the driest ten-year periods in the instrumental record for the area: 1942-1951, 1955-1964, and 1968-1977. During these decadal periods, precipitation amounted to only 73, 82, and 83 percent of average.

Groundwater provided distinct advantages over surface water: it was available year-round and was less sensitive to climate variability (Abruzzi 1985). It also facilitated economic growth and development in the basin, including construction of three coal-powered generating stations (near St Johns, Springerville, and Joseph City), a paper mill, and a mining operation. These industrial activities increased demand pressures on available water resources and generated tensions—which still persist—among local residents. Increased appropriation of water combined with drought conditions prompted the state of Arizona to declare a "critical groundwater area" near Joseph City in 1948. The designation, formalized as the Joseph City Irrigation Non-Expansion Area in the state's 1980 Groundwater Management Act, was introduced as an effort to curtail declines in groundwater supplies through prohibiting expansion of irrigated acreage.

The groundwater basin underlying the Joseph City portion of the watershed came under particularly intense pressure beginning in 1960 due to substantial growth in water demand emanating from construction and expansion of the Cholla power generating plant. The Arizona Public Service (APS) utility, which serves the greater Phoenix area, had bought land or water rights within a 16.5-mile radius of the plant; its purchases included all of the wells except two that were owned by the Joseph City Irrigation Company. These purchases plus designation of the groundwater basin as the Joseph City INA served to preclude further competition between irrigation and power plant water uses. Today, use of water for irrigation continues to be restricted to acreage that had been irrigated during the years from 1975 through 1979 (ADWR 2006).

The amount of water pumped by industry is not insignificant: the Cholla generating station used around 7,425 acre-feet of water in 1980; the paper and pulp mill constructed near Snowflake was using 16,469 acre-feet in the same year. At the same time, however, the amount of water extracted for agricultural uses declined in tandem with reduction in the amount of farmland being actively irrigated. Today, although irrigated agriculture continues to use the greatest amount of water in the region, irrigated acreage has decreased. For example, in Navajo County, irrigated farmland decreased from 11,390 acres in 1978 to 6,351 acres in 1992. Nevertheless, agriculture today accounts for less than 2 percent of the region's total employment, a sharp contrast to the

40 percent thus employed in 1940. The fastest growing water use, by contrast, is in industrial and export activities.

Whereas the drought of 1955-1964 and dry conditions between 1968 and 1977 (Ni et al. 2002) placed pressures on the growing and sometimes conflicting water demands within the Little Colorado River watershed, the wettest conditions in the instrumental record occurred on the heels of the latter drought period. It was during this period, beginning in 1978, that adjudication of water rights began within the watershed (ADWR 1989). It is notable that the 1980 Groundwater Management Act was enacted during this period. The act emerged in response to federal demands that the state get its groundwater overdraft problems under control before approval would be granted for construction of the Central Arizona Project (CAP), which would deliver Colorado River water to the state's main cities and the heavily agricultural area in Pinal County (located between Phoenix and Tucson). While strong regulation of groundwater applied only to the central corridor of the state, the act did address rural issues to a minor degree with establishment of state-level administration, through the Arizona Department of Water Resources (ADWR), of surface-water rights as well as responsibility for collecting and archiving data about water resources in the state and regulating the spacing of groundwater wells.

A proposed interbasin transfer to support mining-related operations outside the Little Colorado watershed illustrates one hotly contested aspect of contemporary water politics in Arizona. In a 1992 decision, the special master of the Arizona General Stream Adjudication addressed a dispute over a proposal by Phelps Dodge to divert water to which it had rights in Show Low Lake to its operations in the Salt River and Gila River watersheds. The special master mandated that "when water is diverted into another watershed or river system as part of an exchange agreement, DWR shall report information on all aspects of the exchange sufficient to allow potential objectors in the watershed of origin to make an informed decision about whether or not to object to the claimed water rights" (Office of the Special Master 2003). The decision illustrates the limitations of state authority over local water resource management: DWR is required to supply information, but the actual decision on whether a water transfer can occur is left to the parties involved. In this case, the dispute became moot when, in 2005, Phelps Dodge Corporation relinquished its rights to Show Low Lake.

A Contemporary Environmental History of the Watershed

Access to—and wise use of—water resources is key to sustainable habitation in the Upper Little Colorado watershed. Climatic variability and change, combined with significant changes in population dynamics and related resource use over the past several decades, underlie water resource vulnerability in the area. Significant growth in tourism and recreation activities, for example, draws large numbers of people to the area. Though the area is experiencing greater winter tourism and recreation such as siding, the area continues to host considerable numbers of visitors during spring, summer, and fall, when water resources may be most stressed. Also significant have been negotiations involving settlement of Indian water rights claims. Climate factors, including a marked warming trend over the past several decades, wet conditions between the 1980s and mid-1990s, and reemergence of drought in the late 1990s and early 2000s, continue to pose challenges to efficient and equitable management of the region's limited water resources. Population growth remained high in the most recent census, showing increases of 4 percent in Winslow, 12 percent in Pinetop-Lakeside, and 22 percent in Show Low between 2000 and 2004. During roughly the same period, between 1999 and 2003, only twenty-three inches of precipitation fell in northeastern Arizona (Morrill 2005). A comparison of this drought with the extended drought period from 1947 to 1956 indicates that, while both periods exhibited significantly below-average precipitation, the latest drought resulted in a consistently lower snowpack. In addition, warmer temperatures combined with the timing of snowfall affected runoff and recharge patterns (Morrill 2005). Research into these conditions suggests that global warming might be implicated in producing more intense drought conditions, in which case greater problems are likely in the future.

Deep, sustained drought exacerbated by climate change may well constitute the gravest threat yet to the sustainability of the human and natural systems in the watershed and beyond. This threat raises serious questions about the future of the area's water resources and demands ongoing adaptation, including identification and enactment of measures to avert or mitigate the impacts of climate variability and change.

Native peoples within the basin have the longest history of interaction with the biophysical environment; however, it is only in the past few decades that they have begun to recapture control over certain important portions of their resource base. Native peoples, in contrast to the other cultural groups in the watershed, place a high spiritual value on water bodies and water resources, a value that embodies the unique environmental histories of the different groups. In the case of the Zuni people, recent developments have led to successful reappropriation of an important water source. The Zuni people had for centuries inhabited land along the Little Colorado and Zuni Rivers, but by 1877, they had lost all but 3 percent of their traditional lands. Most serious was a federal executive order that forced tribal members onto a 408,000-acre reservation in western New Mexico. This forced migration deprived the Zunis of the right to live near the Colorado River, thus disrupting their spiritual practices. Since at least the sixteenth century, the Zuni people have traveled to a sacred location known as Zuni Heaven, a lake located near the confluence of the Little Colorado and Zuni Rivers that has now dried up due to upstream water diversions and groundwater extractions (Committee on Indian Affairs 2003).

Years of effort to reacquire the lake finally culminated, in 1984, in establishment of the Zuni Heaven Reservation for the primary purpose of furthering "the religious and cultural needs of the Tribe" (Committee on Indian Affairs 2002). Once again in possession of their spiritual lands, the tribe began restoring the riparian areas around the lake; in 1990, Congress contributed to their efforts through establishment of the Zuni Heaven Resource Development Trust Fund, which provided $25 million to help restore the area's wetlands. Further assistance materialized in the Water Rights Settlement Agreement of 2003, which empowered the tribe, for purposes of restoring the wetlands, to purchase up to 3,600 acre-feet of surface water rights from upstream sellers and to pump up to 1,500 acre-feet per year of groundwater (Committee on Indian Affairs 2003). Mormon settlers relied on a different set of spiritual values to link their communities with the land and its resources. For example, frequent dam failures and fluctuations in the length of growing seasons, along with the resulting variability in crop output, prompted reliance on religious tenets that viewed natural events such as flooding as trials or tests sent by God and adherence to notions of communal work as religious ritual to improve upon nature. The Mormons' religion-based decision making in the face of environmental stress often involved negotiations between communities that were competing for the same resources. For example, residents of Joseph City (then called St. Joseph) invoked notions of religious community to compel upstream residents of Woodruff to breach their dam. Even though this act could have compromised their ability to grow their crops, Woodruff complied, though they observed that "after working so hard and under such adverse circumstances this was pretty severe medicine" (Dean n.d.). The early reliance on religious rhetoric has since faded, and Mormons no longer make up the dominant Euro-American presence in the watershed. They do, however, continue to represent a locally strong presence in communities such as Snowflake and Joseph City; in locations such as these, religion continues to influence resource management decisions.

The history of the Little Colorado River Basin is one of oscillating periods of plenty and scarcity, with climate constituting a significant but far from determinant influence over the distribution of vulnerabilities to floods and drought. Climate-society interactions in the watershed have strongly influenced the area's biophysical and human environments and have prompted successive innovation and revision of institutional mechanisms aimed at sustaining lives and livelihoods. For the early Mormon settlers, decision making resided at the community level where the main concerns were sustaining the settlements through agricultural production.

Organization of the settlements into stakes, while primarily serving religious and community-building functions, also served as venues for addressing problems generated by environmental variability and change, including variability in hydrologic processes and water availability. Successive droughts, however, eventually resulted in a transfer of decision making from the community at large to irrigation companies, which focused more on rational water management practices than on religious ideals. Throughout the first half of the twentieth century, these companies increasingly turned to groundwater to buffer drought-induced variability in surface-water availability and to meet the considerable growth in water demand. Increases in groundwater use, however, produced overdraft conditions in the area's aquifers, most notably those in the Joseph City area. Here, water table decline induced conflict between power-generation and agricultural uses, eventually leading—for the first time—to substantive state-level intervention and regulation through creation of the Joseph City Irrigation Non-Expansion Area.

The Little Colorado watershed faces an uncertain future characterized by growth in the number of groups wing for access to water and the possibility of greater climatic variability, long-term warming trends, and deeper and longer-lasting droughts. Under such conditions, state-level regulation is likely to increase. A report summarizing an evaluation of the ADWR by the Arizona auditor general has already called for state-level regulation of water throughout the state (Norton 1999).

Knowledge of the environmental history of the watershed provides valuable insight into the experiences, values, knowledge, and practices that have produced the conditions existing today. These "thick" narratives (Geertz 1973), if integrated into development of plans, decision processes, and decision tools, hold promise for improving citizens' acceptance of and support for policies aimed at ensuring the sustainability of the watershed's closely coupled human and natural systems. Scientific research aimed at understanding the nature, intensity, and potential impacts of climate variability and change, and at developing products designed to support policy and decision processes, can also benefit from understanding the contexts in which the products and knowledge will be used.

5

Scarcity, Marketing and Privatization

Turn on the tap and a limitless quantity of high quality water flows for less money than it costs for cable television or a cell phone. The current drought has raised awareness of water scarcity, but most proposals for dealing with drought involve quick fixes-short-term palliatives, such as bans on washing cars or watering lawns except on alternate days. It is assumed that things will return to normal, and we will be able to wash our cars whenever we wish. But the nation's water supply is not inexhaustible. A just-released report of a White House subcommittee ominously begins: "Does the United States have enough water? We do not know." In a survey of states conducted by the U.S. General Accounting Office, only 14 states reported that they did not expect to suffer water shortages in the next 10 years. Is the sky falling? Not yet, but the United States is heading toward a water scarcity crisis: our current water use practices are unsustainable, and environmental factors threaten a water supply heavily burdened by increased demand. As the demand for water outstrips the supply, the stage is set for what Jared Diamond would call a collapse. How will we respond?

When we needed more water in the past, we built a dam, dug a canal, or drilled a well. With some exceptions, these options are no longer viable due to a paucity of sites, dwindling supplies, escalating costs, and environmental objections. Instead, we are entering an era in which demand for new water will be satisfied by reallocating and conserving existing sources. The current water rights structure is the outcome of historical forces that conferred great wealth and power along with the water. The solution to tomorrow's water shortages will require creative answers to challenging issues of equity, community, and economics.

Supply and Demand

In 2000, Americans used a staggering 408 billion gallons of water each day. In many parts of the country, fresh water reserves have been depleted; diversions have dried up rivers and pumping has exhausted aquifers. Industrial solvents have contaminated thousands of groundwater basins, and ocean water has percolated into countless coastal aquifers, rendering them too saline for human consumption. We still have an abundance of potable groundwater, but we are pumping it faster than Mother Nature replenishes it. Additional diversions from our rivers and streams would come at a high environmental cost. In short, our existing use of water is unsustainable.

Moreover, climatic factors threaten the water supply. The recent drought, of historic proportions in some sections of the country, has caused cities, farms, and mining companies to scramble in search of new sources. And global climate change threatens profound (though currently uncertain) implications for the world's water. Notwithstanding the rantings of Fox News, credible scientists no longer doubt the reality of global warming. The release of carbon dioxide gases, a by-product of fossil fuel use, increases the ability of the sun's rays to penetrate our atmosphere, thus raising the earth's temperature.8 Higher temperatures produce a shorter snow season (more precipitation falls in the form of rain), faster snow melt, and increased runoff. These changes have significant implications for our water supply. Think of a mountain's snow pack as a gigantic water storage reservoir. Global warming reduces the amount of water in the reservoir, creating a need to find an alternative means of storage. Global warming also creates higher evaporation losses from the surfaces of lakes, reservoirs, and rivers. According to one recent report, global warming may reduce the Colorado River's reservoir level by one-third by mid-century.

Demands on our water resources are increasing. Increased demands result from one simple fact: population growth. Since 2000, the population of the United States has surged from 285 to 295 million, with the Southwest leading the way. Demographers expect California alone to add 400,000 new residents per year-increasing its population from 36 million to 51 million by 2040. And that number does not include undocumented immigrants, whose current population is estimated at 2.4 million and growing by approximately 180,000 each year. To put this population growth in perspective, consider that California adds one new resident per minute, and that California's rate of growth lags behind that of both Nevada and Arizona.

Fights over water in the United States, no longer confined to the West, illustrate the ramifications of this country's increasing demands for water. The Great Lakes contain 20% of all the fresh water on earth, yet Canada and the eight Great Lakes states have negotiated Annex 2001, an agreement to prevent the bulk exportation of water from the Great Lakes. Bottled water companies have incurred the wrath of citizens' groups in Wisconsin, Michigan, and New Hampshire. Maryland, Virginia, and Washington, D.C. have crossed swords over the Potomac River. New York, New Jersey, Delaware, and Pennsylvania entered an interstate compact that created the Delaware River Basin Commission to manage and allocate the waters of the Delaware River. When municipalities in Massachusetts sought additional water for burgeoning growth, they faced challenges from environmental organizations. Even the humid Southeast is not immune: Georgia, Alabama, and Florida are squabbling over the Apalachicola-Chattahoochee-Flint River Basin. Virginia and North Carolina are struggling over the Roanoke River. Tennessee and South Carolina are trying to prevent the city of Atlanta from diverting water from the Tennessee and Savannah Rivers. Meanwhile, South Carolina faces a threat as North Carolina has taken water from the Pee Dee River for power plants, other industrial users, and Myrtle Beach's half-million people.

Options

As the nation's water use spirals upward, where will the water come from to satisfy new demands? We have five options. First, we could simply continue to exploit the resource by diverting more water from rivers and pumping additional water from underground aquifers. But in many sections of the country our current water use is unsustainable. Our diversions of surface water have completely dried up some rivers and reduced the flow in others to a trickle. It was once thought that water left in a river was wasted, but we now know-thanks to the development of the fields of biohydrology and ecohydrology-that fresh water flows are critical to the survival of riverine species and spawning habitats, and that they allow for intermittent floods which scour sediment, nourish habitat, and impede the encroachment of invasive species. Additional diversions of surface water would come at a high cost to the environment.

Groundwater was once thought to be as inexhaustible as the air we breathe. We now know better. In the 19th century, when the science of hydrology was in its infancy, and American judges were perplexed as to how to divvy up rights to pump groundwater, they threw up their hands and proclaimed, "If you can get it out of the ground, it's yours." And then the free-for-all began. Since then, the science of hydrology has matured into a sophisticated science, but the legal rules have not kept pace. In most American states, groundwater law is governed by the "reasonable use" doctrine-an oxymoron-that allows a person to pump as much groundwater as she desires so long as it is used beneficially. Technological breakthroughs in the 1940s and 1950s vastly expanded the capacity to pump huge quantities of water from extraordinarily deep areas beneath the earth's crust. Large-scale commercial irrigation wells can now pump thousands of gallons per minute. Because the legal rules were so permissive, farmers installed millions of wells and began pumping feverishly. The net effect was a disaster.

Think of an aquifer as a giant milkshake glass, and think of each well as a straw in the glass. The water in the glass is limited, but access to it is not. The reasonable use doctrine epitomizes what Garrett Hardin called "the tragedy of the commons"-limitless access to a finite resource. Anyone can pump as much water as he wishes. There is no incentive to husband groundwater because pumpers do not own the resource. Instead, the system encourages willy-nilly development. In many sections of the country groundwater tables are plummeting. Excessive groundwater pumping has caused the ground to collapse; rivers, lakes, and springs to dry up; and riparian habitat to die. If we continue to exploit our groundwater resources in this way we will eventually run out of water. In the meantime we face higher costs, a decrease in water quality, salt-water intrusion into coastal aquifers, land subsidence, and further environmental degradation.

The second option for satisfying new water demands is expanding our available supply. Dams function essentially as storage reservoirs which ensure that water is available at a time when Mother Nature does not. In the American West, most water comes from winter snowfall in the mountains. During the spring thaw, the snowmelt creates cascading rivers that provide more water than farmers or cities need for that season. During the summer, when the water needs of farmers and cities increase, dams function to smooth out the supply.

Beginning with the construction of Hoover Dam in the 1930s, the American West embarked on a dam-building frenzy. The remarkable construction of Hoover Dam symbolized how the United States could do anything, even harness the mighty Colorado River. To a nation in the throes of the Great Depression, it was a welcome boost, personifying an ambition that promised a bright economic future. And the power for that future would come from dams, lots of them, each producing hydroelectric energy. The water aided western farmers, and the power enabled American companies, such as Boeing, Lockheed, and Martin Marietta, to become aerospace giants. By the end of the dam-building era in the 1960s, most major western rivers had been dammed, some repeatedly. Indeed, in the United States as a whole, there are 75,000 dams six feet or higher and as many as 2.5 million smaller dams. Fewer than 60 American rivers are free-flowing.

In the mid-20th century, it seemed inconceivable that there might be a downside to building dams. But there was. Decades later, we came to realize how profoundly dams alter watersheds. Dam construction inundated some of the most beautiful canyons in the West, such as Hetch Hetchy and Glen Canyon, and transformed the rivers below the dams. Water flowing from a dam has a constant temperature as opposed to one that fluctuates with the seasons. Most dams increase water temperatures, but some decrease temperatures as they release cold water from the bottom of the reservoir. The flow itself depends on the decisions of engineers, not on a natural rhythm. Native fish and other aquatic species suffer from these changes when the nutrients that formerly sustained the downstream aquatic habitat become trapped in the quiet lakes upstream. Anadromous fish-Pacific salmon and steelhead on one coast, and Atlantic salmon and shad on the other-found the paths to their spawning grounds blocked by impassible edifices. Some species became extinct; others merely suffered. Dams also decrease the level of oxygen in reservoir water. The release of this oxygen-deprived water can kill fish downstream.

Today, opposition to building new dams is quite substantial. Having witnessed the profound alteration of the hydrologic regime of dammed rivers, the environmental community is adamant that new dams are not an environmentally acceptable solution to problems of water shortage. Apart from environmental concerns, dams face two other obstacles: enormous costs and a paucity of good dam sites. Therefore, both option one-diverting more water from rivers-and option two-building new dams-prove to be problematic alternatives for increasing water supplies.

A few smaller dams are still being built, such as the controversial Animus La Plata in Colorado, but the movement in the United States is in the opposite direction. We have begun to decommission dams. Since Edwards Dam in Maine was taken out in 1999, 140 other dams have been removed, restoring miles of free-flowing rivers. There is traction to this movement, as communities have realized the considerable economic and environmental value in removing aging dams that generate little hydropower. And it is not just small dams on small rivers that are being targeted. A major debate is raging about removing four substantial dams on the Snake River that block salmon migration. The year 2004 saw a renewed focus on restoring Hetch Hetchy Valley by removing the dam. And Glen Canyon Dam on the Colorado River is under attack as wasteful and environmentally destructive.

Are there other options to expand our water supply? Desalination of ocean water offers the prospect of solving the lament of Samuel Taylor Coleridge's ancient mariner: "Water, water, everywhere" but not "a drop to drink." Removing salt from ocean water to make it potable offers a tantalizing possibility of an abundant new source of water. Some middle eastern countries and Caribbean Islands have obtained drinking water from desalination plants for a long time. In 2001, Tampa Bay Water began construction of the largest desalination plant in the Western Hemisphere. The $110 million plant came online in March 2003 and was expected to produce 25 million gallons of potable water per day. Alas, the plant only operated for two weeks before problems cropped up. The intake pipes sucked in Asian green mussels-an invasive species creating headaches up and down the East Coast. Sediments quickly clogged the expensive membranes, requiring their immediate replacement. In November 2004, amidst one bankruptcy and a flurry of lawsuits, Tampa Bay Water agreed to pay a California engineering firm an additional $29 million to fix the pretreatment system. These troubles prompted the board chairman of Tampa Bay Water to note wryly: "[B]eing on the cutting edge is not a very comfortable position...."

Even if Tampa Bay Water resolves these problems, desalination must surmount other problems before it becomes a widely-available, cost-effective, environmentally-friendly source of potable water. The desalination process incurs large costs due to the high-tech membranes used to filter out the salt and the immense amount of energy required to run the plant. Also, various methods of desalination, including reverse osmosis and nanofiltration, generate streams of waste: every 100 gallons of seawater will yield between 15 and 50 gallons of potable water, leaving 50 to 85 gallons as a super-saline byproduct. Disposing of this brine poses significant engineering and environmental challenges. In Florida, fishery ecologists fear that the release of this super-saline water into sensitive estuaries off the West Coast will adversely impact the reproduction of clams and other marine organisms. In California, the Schwarzenegger Administration is enthusiastic about desalination, but the California Coastal Commission is quite concerned about the environmental consequences. The March 2004 report of the California Coastal Commission notes: "Seawater is not just water, but habitat. It provides the matrix within which innumerable organisms live, and serves a critical role in everything from the food web to the climate." The discharge from a desalination plant may double the normal salinity level for local seawater. Such a sharp spike in salinity may kill or have sublethal effects on marine species and other organisms, particularly in the egg, larvae, or juvenile life stages. Finally, the Coastal Commission report raises a basic question about desalination projects proposed by private, for-profit corporations: "Should seawater, a public resource held in common for the benefit of current and future generations, be allowed to be expropriated by private business for profit?" Desalinized water has a current place in the water portfolios of some American municipalities. But it is an emerging technology-not a quick-fix solution to the problem of water scarcity.

The third option for satisfying new water demands, one that is technically viable today, is the reuse of municipal effluent. Historically, cities dumped raw sewage into our rivers. Today, most household wastewater, whether from toilets or showers, ends up at the municipal wastewater treatment facility. Until recently, this facility would filter the water under the standards mandated by the federal Clean Water Act and then discharge it into a nearby river. From the city's perspective, the point was to get rid of it as cheaply as possible without generating litigation or complaints about odors. Now, ironically, effluent has economic value. Many industrial users realize that effluent provides a perfectly adequate water supply, joining farmers who have used effluent for crops and cities that have used it for golf courses, municipal parks, cemeteries, and roadway medians. Indeed, wastewater treatment technology can take sewer water and clean it up to drinking water quality, though most Americans would rather not dwell on this prospect.

In 1998, San Diego floated a trial balloon along these lines, but the program-dubbed, by some, the "Toilet to Tap" proposal-was dead on arrival. Squeamishness aside, astronauts have lived with total water-recycling programs since the beginning of the space program. In any event, using municipal effluent for purposes other than drinking water would expand our water supply. Yet the treatment process is quite expensive and delivery of effluent for nonpotable purposes requires a completely separate set of pipes and valves-a daunting financial prospect for cash-strapped American cities. Effluent reuse will help, but not solve, the problem of finding an adequate quantity of water to supply the increasing demands of a larger population. As a fourth option, we could make our existing supply last longer by using water wisely and efficiently. The impetus for water conservation might come from two directions: government rules and regulations or market-based price signals.

An example of government rules and regulations is Arizona's Groundwater Management Act, which imposes on all users-cities, farms, and mines-conservation standards that are phased in over 45 years. In each successive ten-year period, the standards tighten like a ratchet, requiring more effort from the users to comply. After 25 years of experience, the results are mixed. Some progress has been made, especially in communities like Tucson that have an ethic of water conservation.

Public service messages and programs reinforce the idea of contributing to the community's best interest by "Beating the Peak" usage in the city's hot summer days. Yet Tucson's population continues to swell, and, while it is nice to boast that per capita per day consumption rates have declined, the higher population has resulted in increased total consumption. Just as it is easier to pick low-hanging fruit, Arizona achieved initial success with nobrainer programs that targeted obvious waste, such as apartment complexes whose sprinkler systems sprayed as much water on streets as on lawns and shrubs. Now the State is in the later stage of reaching for the higher fruit. The cornerstone of the Act is a program that requires developers to demonstrate an "assured water supply" before getting their subdivision plats approved. Developing regulations to implement that program has been vexing for the

Arizona Department of Water Resources. Countless public meetings and innumerable drafts-with debates over every comma and semicolon-finally yielded rules totaling 36 pages of single-spaced fine print. Conservation standards fraught with complexity, thereby requiring elaborate monitoring programs, may prove to achieve neither cost effectiveness nor meaningful results.

Still, state and local governments have a critical role to play in developing water conservation programs. One spectacularly successful program has created financial incentives for homeowners to replace toilets and showerheads with low-flow fixtures. Water conservation standards for new development are even easier and cheaper to implement than incentives for retrofitting existing structures. Housing developers do not care whether they instruct their plumbers to install toilets that use 1.6 gallons or 6.0 gallons per flush. Peter Gleick has recently demonstrated the tremendous potential for water conservation programs in California to help alleviate water shortage problems. He calculates that if the state funded a program to replace all remaining high-water-use toilets with low-flow ones, it would save 410,000 acre-feet of water per year. This is serious savings.

An alternative water conservation strategy would gain people's attention about their water use through their pocketbooks. If the price of water rose, people would carefully examine how they use water, for what purposes, and in what quantity. Alas, the price of water in the United States is ridiculously low. Many Americans pay more for their cell phones and cable television each month than they pay for water. Whether a consumer receives water from a municipal water department or from a private company regulated by the state public utility commission, the bill that she receives is only for the "cost of service"-those costs associated with delivering the water. There is no commodity charge for the water; it is literally free because water departments and utilities do not pay for the water themselves, and they pass that benefit along to consumers. This will change when cities are forced to acquire new supplies by entering the market to purchase water rights from willing sellers at market rates. But for the moment, existing water bills do not reflect these new costs.

Indeed, residents of some cities do not even pay for the cost of delivering the water. The municipality, as a service, provides water to all residents who may use as much water as they wish. Some communities do not have residential water meters, and millions of apartment complexes lack individual meters for each apartment. In Fresno, California, a controversy erupted in 2003 as to whether meters should be installed in people's homes. In the end, the state legislature passed a bill requiring the installation of meters-by the year 2025. Meters enable a city to insist that residents be responsible in their water use, or pay financial consequences. The absence of meters has significance for water use. In Fresno, a city without meters, per capita use hovers around 300 gallons per day; in neighbouring Clovis, which has meters, use is approximately 200 gallons per day.

Water rates are politically controversial, yet we must begin to confront the perversity of encouraging wasteful water use. Presently, over half of California's water providers use a flat rate or a declining block-rate structure, which rewards the highest water users with the lowest rates. We must reverse that. Sensible water policy should create incentives to conserve by imposing an inverted block-rate structure that targets the heaviest consumers and makes them pay considerably more for their excessive use. The heaviest residential consumers use a disproportionate percentage of their water outside the home for swimming pools and lush landscaping, two quite discretionary uses. As we raise water rates, we must be sensitive to the impact on families of modest means. A threshold that eliminates, for example, the first 5,000 gallons each month from any charge would address this question. Targeting rates based on the differential between winter and summer usage would work as well. Those consumers whose water consumption jumps 50% between the seasons are using water for lawns and swimming pools. People do not use more water for toilets, showers, or washing machines in the summer than they do in the winter. In short, water conservation standards and appropriate water rates offer significant opportunities for stretching our current water supplies.

Desalination, effluent reuse, water conservation, and water pricing will help secure additional supplies and reduce demand in the future. But the pressure to find more water continues. Where will the water come from? We cannot make more water because the hydrologic cycle is a closed one. That leaves a fifth option for satisfying new demands: we can reallocate water from current uses to new ones. We are about to enter an era of water reallocation. How will this reallocation take place? The government could mandate transfers from one user to another, except that would generate bitter political controversy and litigation over whether the government has the authority to act so cavalierly and whether the Constitution prevents the confiscation of water rights.

The Promise and Prospects of Water Markets

It would be far better to encourage voluntary transfers between willing sellers and buyers. Let them decide what the water is worth to each of them. Water markets would facilitate the movement of water from low-value activities to higher-value ones, thus resulting in a more efficient deployment of the resource. In the United States, we waste an immense amount of water growing cotton and alfalfa to feed cattle. In California, farmers grow cotton on 750,000 acres, heavily subsidized by the federal government. Irrigation systems are often primitive earthen canals that lose 40% to 50% of the water diverted into them through seepage into the ground. Once the water arrives at a farm, many farmers use highly inefficient flood irrigation or sprinkler irrigation with nozzles directing the water into the air where much of it evaporates. By one measurement, almost one million acre-feet of the three million acre-feet diverted by the Imperial Irrigation District in southern California ends up as wastewater flowing into the Salton Sea. In the United States, farmers irrigate millions of acres of marginal land, not because they produce high yields or generate substantial profits-in fact they do not-but because the fanner has the right to irrigate. Indeed, failure to do so may result in losing the water right through the doctrines of abandonment or forfeiture. Although some farmers have adopted water-saving technologies, agricultural subsidies and water subsidies combine to distort the economics of agricultural production. As a result, many farmers persist in growing low-value crops and fail to reduce the waste of artificially devalued water resources.

This byzantine system needs a major overhaul. Part of this reform should come from government rules and regulations that impose conservation requirements, eliminate subsidies, encourage investment in modernization, and require "full-cost pricing"-the beneficiaries of U.S. Bureau of Reclamation projects should pay the actual cost of the water they receive. But these changes, each desirable in the abstract, would be extraordinarily difficult to execute in the concrete. Several involve very expensive system improvements, such as lining canals with concrete or laser-leveling fields, that can cost hundreds of thousands of dollars for a single farm. Where is this money to come from, given that many farmers operate on razor-thin margins? Because the price of food relative to inflation has remained stable since the 1940s, it is a wonder any farmers can make a go of it. As a sardonic expression about making money in farming puts it, "If you want to make a small fortune in farming, start with a large one." It is not feasible, reasonable, or equitable to require farmers to undertake massive expenditures in order to make their irrigation systems more efficient. Even if it were reasonable, it will not happen for one very practical reason: farmers yield immense political power. State legislators would act at their peril were they to require their farmer constituents to shoulder the burden of these huge expenses. As the doctrine of public choice instructs, politicians like to remain in office.

Perhaps other constituencies could pressure legislators to impose conservation standards on farmers. The rise of the environmental movement, the growing demands of cities, and the increasing role of recreation on public lands in the West has created groups of voters whose interests in water differ from those of farmers. Even though these constituencies consist of a large number of voters, they are geographically diverse and lack a focus on any particular issue involving farming practices. For example, a politician contemplating how to vote on a bill to require farmers to laser-level their fields would be confident that 100% of the farmers would be strident opponents, but he would be uncertain as to the views of his other constituents. It is the phenomenon of a small number of deeply-committed voters yielding inordinate political influence over the legislature. Politicians listen carefully to the dominant economic interests in the state. Recognizing this political reality, we must accept the fact that state legislators will not impose costly changes on the farming community. The best way to reform agricultural water use in the United States is to give farmers a financial incentive to use less: let them sell water to cities.

Market-based transfers can take many forms, from sales to leases, from forbearance agreements to dry-year options, and from land fallowing to conservation measures that save water. Each offers the prospect of a win-win result as the seller secures a price that she finds attractive and the buyer secures a water supply worth the negotiated price. The case for water marketing rests on the assumption that ownership of an item invests the owner with an incentive to take care of it. While this is surely not a universal proposition-as human beings span a range from the most obsessive-compulsive among us to those who seem oblivious to disarray, and from those who find it difficult to part with a nickel to those who cannot seem to hold on to one-ownership still changes behaviour. Consider rental cars and ask yourself whether you would treat a new car that you purchased the same way you treat a rental car. When, for example, was the last time you washed a rental car? The same point might be made about hotel rooms, public parks, and parking lots. I have never seen a cigarette smoker dump his ashtray out in his driveway, but some smokers do not hesitate to do so in parking lots. That is not to say that all people treat public property recklessly. Some of us pick up trash in parks and on hiking trails, while others tidy up hotel rooms. Whether driven by Kant's categorical imperative, inner feelings of guilt or shame, concern for chambermaids, or a belief that civility ennobles us as a people and a culture, many Americans take responsibility for public places. Others, alas, habitually trash public spaces. Whether the habit of littering is cultural or rooted in status and class, people act differently toward things they care about.

An ability to transfer ownership creates an incentive to use property more productively. This is the core idea of markets. Owners of property assess the value of it to them and part with it if they will realize a profit. Buyers seek to change the use of property and capture the value added by the new use. In this process, both sellers and buyers make profits, and society benefits from increased efficiency.

Water markets have other benefits. They permit the reallocation of water in response to changes in population and economic development. To take one example, the computer industry in California's Silicon Valley has transformed the American economy and required that we find water for those who work in the information technology field and for the industries that make the chips and routers that fuel the web. In California, 1,000 acre-feet of water generates 9,000 jobs in the semiconductor industry but only 3 jobs producing cotton. Each acre-foot used by the semiconductor industry produces nearly $1 million in gross state revenue but only $60 growing cotton and alfalfa. At the same time, NAFTA and other international trade agreements have opened American markets to agricultural powerhouses in Latin America and Asia, putting pressure on our farmers who have watched their margins erode. The time is ripe for some farmers to transition out of growing crops whose economic yield does not warrant the time, effort, and money that it takes to grow them. Water markets provide them this option.

Water markets may even encourage water conservation. If a farmer who reduces water loss by lining his ditch with concrete can profit from the sale of water he has conserved, it stimulates the investment in conservation practices and frees up the water saved for other users. The environmental community has also come to recognize the potential benefits of water markets. The transfer of water from farms to cities lessens the pressure to build new dams, to divert even more surface water, and to pump more groundwater. And it has allowed some environmental groups to purchase water rights from farmers that they then dedicate to in-stream flow rights, thus ensuring minimum flow levels in sensitive rivers and streams.

Let's be clear about one thing: we are talking about transfers from rural farming areas to cities. Most of the water that will sustain the expected 15 million additional Californians is going to come from agriculture. It has to. In California, as in most western states, farmers use between 70% and 80% of the state's fresh water. One cannot seriously address the question of new demands for water without focusing on agriculture. Another driving factor is money. In many western states, a high percentage of agricultural water is used to grow cotton and alfalfa, crops that return a relatively low value. The economic value of this water to cities dwarfs the value of the same water to the farmers. It makes economic sense to let the water support the higher value activity. If water markets are to flourish, there must be a system of quantified water rights that are transferable. Water markets can only develop if a farmer has a known and fixed right that she can sell or lease. Without a property right that is quantified and transferable, there will be no voluntary reallocation of water use.

Community Impacts and Equity Considerations

The state should not give unrestricted permission to transfer water. Water is a public good and a public resource. The transfer of water from agriculture to cities will benefit both farmers and urban interests, but it may harm third parties. Communities have developed around agricultural centres to serve the farms and to provide a pool of labour to maintain the agricultural economy. If a farmer sells water to a city, his decision has implications for his farm workers; his John Deere dealer; his pesticide, fertilizer, and seed suppliers; and his lawyer, accountant, and banker. Also affected are local restaurants, supermarkets, and retailers who provide necessities and small luxuries to low-wage agricultural workers. Even local government will suffer financially from lost or lowered property values, sales, and income taxes, and from the increased need to provide social services to displaced workers. Equity demands that the beneficiaries of the water transfer compensate all those hurt by it.

Government must oversee the transfer process by setting standards regarding who is entitled to compensation, for what, and for how long. The trick will be to ensure fair compensation for those harmed by the transfer without creating a cumbersome hearing and appeal process that would drive up transaction costs and hijack the transfer process.

Government must also ensure that environmental factors receive careful consideration. Market systems have difficulty internalizing environmental costs. Economists expect that a rational owner of private property will protect the environment on his own property. But a water transfer may affect the habitat on someone else's property, such as the land of a downstream neighbour or a state wildlife refuge.

For water transfers to become a legitimate tool for water reallocation, they must internalize both third-party and environmental costs. Even then, government may occasionally prohibit water transfers in order to protect valued and unique communities. For example, northern New Mexico's acequias are centuries-old subsistence-farming communities of Hispanic Roman Catholics that conceive of water as a community resource. The State of New Mexico has a compelling interest in protecting this rich culture's traditional water use.

Before we rush headlong into water marketing, we must confront a deceptively simple question: Can someone sell water? Is water a public resource, essential for life, not a commodity to be bought and sold like porkbelly futures? In many societies water has spiritual, religious, and cultural aspects. To conceive of water as private property-owned by someone who can unilaterally decide whether to sell it, to whom, for how much, and for what purpose-raises profound philosophical and moral issues as well as troubling political questions about the role of corporations, especially multinational corporations, and about the ability of local communities to be independent, autonomous, self-sufficient, and self-determinative. To critics of privatization, a society is bankrupt of values if it treats water as simply a marketable commodity, no different than video games or kitchen faucets.

Privatization of Water

International controversies over water privatization are shaping the debate in the United States. In a world of six billion people, where over one billion lack access to safe, potable, and affordable water, the issue of privatization of water resources poses an immense challenge to the international community. The context is etched sharply by recent strife in Cochabamba, Bolivia. In 1998, the World Bank insisted that the Bolivian government turn over its public water utility to the private sector, or else the Bank would refuse to guarantee a $25 million loan for improvement of the water system infrastructure. The Bank required that infrastructure costs be passed on to consumers. At the instruction of the Bank, the company that received the concession, a subsidiary of the Bechtel Corporation, increased water rates by 35%. A series of escalating protests resulted in seven deaths and spurred Bolivian President Hugo Banzer to place the country under martial law. Other demonstrations have occurred in Argentina, Ecuador, Panama, and South Africa.

To many progressives, these incidents throughout the Third World share one thing in common: multinational corporations exploiting the dire economic situation of poor people. Corrupt political regimes, often bribed by these companies, pay no heed to citizens' complaints. These episodes have led some opponents of privatization, such as Maude Barlow and Tony Clarke, to draw a line in the sand, stating that "[t]he move to commodify depleting global water supplies is wrong-ethically, environmentally, and socially." They think privatization of water resources allows allocation decisions to be made by corporations that desire to maximize profits and ignores the environmental and social consequences of water allocation policies. These companies, focused only on the bottom line, are unlikely to invest in new technology or water conservation.

To Barlow and Clarke, privatization interferes with citizens' ability to allocate and manage their own water, concentrates power in the hands of monopolist corporations, and makes it difficult for local governments to reclaim control over the water system. So is privatization a good thing or a bad thing? As with so many other things in life, it depends. The devil is in the details. To some, what happened in Cochabamba epitomizes what is wrong with privatization. But since the uprising, the cooperative-run water system that replaced Bechtel is in shambles, possessing neither the capital to install or overhaul the infrastructure nor the experience to run a public utility. To analyse the situation in Cochabamba or elsewhere, one must know the state of affairs before the private company arrived. What was the condition of the infrastructure? Was it decayed and neglected? Did everyone in the community receive water before the company came in? And what exactly did the company do? Did it build, repair, or replace the infrastructure; deliver water to people;

charge people for water delivered; respond to the demands of local politicians to divert resources to their pet projects?

How does one judge the profits to be huge? If a company invests tens of millions of dollars in rebuilding a crumbling infrastructure, in restoring and expanding water delivery to poor urban and peri-urban communities, and in putting in place a competent water administration system, it quite justifiably expects the return of its capital and a reasonable profit. So the profit motive alone is not enough to condemn the corporation as exploitative or privatization as a bad idea. The passion generated by water privatization is nicely captured by an exchange between an Argentinean opponent of privatization, who argued that water "is 'a gift from God,'" and the President of Vivendi Environmental (which supplies water to 100 million people throughout Europe, Asia, Africa, and the Americas) who responded, "'Yes... but he forgot to lay the pipes.'"

Despite the intense debate, privatization is an elastic concept that embraces many different scenarios involving the transfer of the assets or operations of a public water system into private hands. Most water systems in the United States are publicly owned and operated. Things were not always this way. In the early 19th century, most citizens received water from a private water company. At the end of the 19th century, municipalities began to assert control over these services because they recognized that private companies were not providing adequate service to all citizens. Private companies often failed to invest sufficient capital in the system and sometimes supplied water to the wealthier sections of a city and not to the poorer sections. Issues of water quantity and quality took second seat to maximizing the company's profit on its investment. By the year 2000, private companies served only 15% of the American public.

Recently, the pendulum has swung back again. The drivers behind water privatization come from several impulses. First, financially strapped municipalities are eager to have a private corporation put forward the huge amount of capital necessary to update the obsolete and decaying infrastructure of municipal water and sewer systems. By one federal estimate, it will require one trillion dollars over the next 20 years to replace aging sewer pipes and treatment plants. Second, many economists argue that private businesses are more cost-efficient and effective in providing services than the public sector. And third, in some quarters ideology favours downsizing government and outsourcing things to the private sector.

There are several forms of privatization, each with different implications for the debate. A limited, often uncontroversial form of privatization involves a local government contracting with a private water company to operate the municipal water system or the wastewater treatment system. Municipalities regularly request bids from the private sector to design, construct, operate, and maintain public facilities. Even though public employee unions might prefer that the jobs remain in the public sphere, few other people would object if private companies were to administer the billing and revenue collection services or the payroll obligations of a municipal water department. Since 1997, the number of publicly-owned water systems operated by private companies under long-term contracts jumped from 400 to approximately 1,100. Even still, 94% of water systems in the United States are publicly controlled.

A more contentious type of privatization involves selling or transferring the assets-the pumping plants, treatment facilities, headquarters buildings, and distribution systems-of the municipal water system to a private company. Often this exchange is the quid pro quo for the company agreeing to infuse the system with a major dose of new capital. In this situation, the municipality avoids the need to rely on its municipal bonding and financing system to generate new monies for the project, which would saddle residents with higher taxes. Whether this form of privatization is good or bad depends on the understanding between the municipality and the private contractor. Does the municipality have adequate oversight on issues of water quantity and quality? Will the company undertake sufficient water conservation efforts, such as attempts to increase the use of municipal effluent? The answers depend on the specifics of the contract between the two parties. The most bitter international controversies have involved situations where local governments have entered into contracts with the private sector to own and operate the system for a substantial period of time, often 50 years. The time period necessary to recoup the heavy initial investment by the private corporation justifies the length of the contract, but the duration may effectively cede control over a public resource to a for-profit corporation.

The final form of privatization, the one that really gets the juices flowing, that most resembles a red flag in front of a bull, involves ownership of the water itself. If a private corporation owns the water, may it sell the water at whatever price it wishes to whomever it wishes? This form of privatization presents a number of problems. Given that government's essential role is to provide basic services and water, perhaps government has defaulted on its responsibility. If companies own water, they may distribute it unequally, favoring the wealthy who can pay more and the politically powerful who can help in other ways. The contract between a government and a private corporation will determine which residents receive water and at what price. Privatization risks shutting out the public from participation and may make a company's practices less transparent. When economic conditions produce natural monopolies of scarce resources, we create regulated industries. Whether this regulation is successful depends on the strength of the political institutions.

That is why the criticism of water privatization in Third World countries does not transfer very well to the U.S. domestic situation. In the Third World, privatization makes the most sense in those places where the government has failed to provide basic human needs for their people. But in those countries with weak, ineffective, or corrupt governments, privatization presents a problem because the governments may not adequately regulate the private sector. The United States, by contrast, has a strong tradition of democratic oversight of private utilities through public utility commissions (PUCs). PUCs exist in every state and the membership usually consists of elected officials beholden to the public. Most PUCs see their mission as placing a strong thumb on the neck of utilities that might otherwise gouge consumers. I am not suggesting that PUCs always provide the most efficient, democratically accountable regulation of private utilities, but they do an adequate job in most instances.

Environmental consequences and water quality are two other issues surrounding privatization. A private corporation has little incentive to be concerned about the environmental impact of providing water. If the local water supply is groundwater that is being used in an unsustainable fashion, the consequences on the environment may be immense. Similarly, surface water diversions may have horrible environmental consequences but be of little concern to the private corporation that diverts the water. These environmental costs are not internalized by the company but shunted off on society generally. As for water quality, private companies often resist undertaking expensive monitoring programs for low levels of pollutants. Corporations fear, often reasonably, that it will be difficult to recoup these costs through rate increases, which are subject to both consumer acceptance and PUC approval.

Some opponents of privatization claim that if a country allows its resources to be privatized, international agreements will require unlimited access to the country's water supplies, thus setting off a mad scramble by foreign corporations to export water for sale. This fear rests on provisions in the General Agreement on Tariffs and Trade83 (GATT) and the 1994 North American Free Trade Agreement (NAFTA). GATT governs international trade for the 148 countries who are members of the World Trade Organization (WTO). Some environmentalists believe that if a country permits bulk transfers of water domestically, GATT would prevent a country from regulating further trade in its water resources. However, there are two exceptions to the bulk transfers requirements: actions by a country to protect human life or health and actions to conserve exhaustible natural resources. Though the legal question remains unresolved, it is likely that these exceptions mean that any country that imposes water use restrictions on its own citizens for purposes of conservation need not fear that GATT will require it to open its water resources to international marketing. As for NAFTA, it places "ordinary natural water of all kinds" under a tariff heading, implying that water can be traded internationally. However, a 1993 joint declaration signed by Canada, the

United States, and Mexico specifies that water is not covered by NAFTA unless it has entered into commerce and become a good or product. "And nothing in NAFTA would obligate any NAFTA party to either exploit its water for commercial use or to begin exporting water in any form."

A practical reason suggests that the fear of transfers between countries or continents and across oceans is more theoretical than practical. Water is heavy: about 2 pounds per quart or 1,358 tons per acre-foot. Wide-eyed dreamers have proposed moving water by tanker from Canada to the Caribbean, by immense water balloons from the Arctic to California, and by ships towing icebergs from the Antarctic to the Middle East, but few investors have placed serious money on the table to underwrite these schemes. The practical reality is that water has such little value as currently priced that these transfers are not economically viable. When the price of water begins to approximate the price of oil which is moved by tankers, it will be time to resolve the GATT and NAFTA issues.

A major weakness in the arguments of the antiprivatization critics is that they have not suggested an alternative. There is really only one: allocation of water through the public sector. Such allocation would be done at the direction of elected politicians or at the discretion of bureaucrats. No economomist thinks that this is an efficient way to make resource allocation decisions. Indeed, that's the problem with the existing distribution of water in the American West. The Bureau of Reclamation has curried favour with important members of Congress and important agricultural organizations and has distributed the water accordingly. Allocation decisions made through the political process will invariably result in the water being allocated to the most powerful economic interests in the state.

As critics of water privatization, Maude Barlow and Tony Clarke argue that water is an inalienable political and social right, and that each person should be guaranteed a "water lifeline," which they calculate as 6.5 gallons per day. Peter Gleick also argues for a human right to water and has analysed that it takes a minimum of 13 gallons of water per day per person for drinking, cooking, bathing, and sanitation. Who could disagree? After all, the United States-the richest country in the history of the world-can easily make this commitment to its people. However, recognizing a human right to water does not resolve the issue of privatization; indeed, it begs the question. The amount of water needed to supply basic human needs constitutes only a tiny fraction of the water used each day in the United States. Thirteen gallons per day multiplied by 295 million people in the United States totals roughly 3.8 billion gallons of water per day. That's less than 1% of the 408 billion gallons per day used for all purposes in 2000. The real issue confronting the United States is not whether to recognize a human right to water, it is how to allocate the remaining 99% that we use each day. So let's recognize a human right to water and focus on how to distribute the rest of the water. The question becomes what is the appropriate role for water privatization with respect to the remaining 99%?

For privatization to be successful, governments must regulate water as a social good, ensuring access to all at a fair price. PUCs must carefully monitor the financial returns to the private company and link any rate increases to agreed-upon improvements in service, conservation programs, or environmental stewardship. Moreover, government should retain ownership of the water resources. In the United States, there is no reason to surrender the ownership of a municipal water supply to a private corporation.

Newcomers versus Existing Users

In the United States, the poster child for what is wrong with the privatization movement is Nestlé Waters North America, a subsidiary of the Swiss-based food giant Nestlé. The company commands a 32% share of the American bottled water market and, in its quest to supply "spring" water to American consumers, controversies over its water use have erupted in Texas, Florida, Maine, Michigan, New Hampshire, Wisconsin, and California. A foreign multinational coming in and bottling millions of gallons of local spring water, only to transport it out of the basin, presents moral, ethical, environmental, and economic questions. I may have been the first to point the finger at Nestlé. Nevertheless, what Nestlé is doing is no different than the actions of many other entrepreneurs.

For example, consider brewers of beer. Whether it is Budweiser, Miller, or Coors, American brewers use huge quantities of local water to brew beer and then ship the beer out of state. The question is whether it is preferable to allow the brewing of beer over the bottling of water. Millions of beer-drinking Americans think this is an easy choice. Nevertheless, both bottled water and bottled beer involve the export of large quantities of local water for human consumption.

There are indirect transfers of water out of basins as well. Consider a processing plant for Del Monte's green beans or canned corn. Each can contains a little bit of water from the local area that is sent out of state. Or consider crops, such as tomatoes or cantaloupes, that consist almost entirely of water. To be sure, the farmer has input both labour and materials to seed, cultivate, and harvest the crops. But, like a bottled water operation, it is a transfer of water from in-basin to out-of-basin.

What distinguishes these situations? For me, it is the distinction between current users and future users. In times past, we allowed Coors to divert huge amounts of Rocky Mountain water, which created problems for downstream communities and the environment. Were we to take a fresh look at whether to permit Coors to open a new brewery, we might decide that the answer is no. But the plant is already in existence and in operation. That distinguishes Coors from a new, highly consumptive use of a public resource that may either interfere with the public water supply or cause undesired environmental harm. We must and should grandfather in existing users, if only for the political reasons noted above, but there is no reason to treat newcomers like Nestlé equally. Such new-entrant water entrepreneurs have no claim to water based on history, community, vested rights, or equity. They should be required to demonstrate that their proposals are consistent with the public interest and will not cause environmental harm.

The law in some states distinguishes between current and new users. The seminal 1980 Arizona Groundwater Management Act mandated two policies that are critical to the reallocation of water from one activity to another. First, it grandfathered in existing users and quantified their rights. Second, it made those rights transferable. This paved the way for market transfers from existing to new uses. The State of Arizona coupled these changes with a program that requires developers to demonstrate that they have an "assured water supply," which is defined as sufficient water for one hundred years, before they can sell homes or lots. Think of my metaphor of the milkshake glass. Before a developer can drill a new groundwater well, the government should insist that the developer pinch someone else's straw. That is, the developer must purchase and retire an existing water user's right in exchange for permission to place a new demand on the aquifer.

Water Rights in the United States

Regardless of the normative arguments as to whether water should be privatized, the legal system in the United States recognizes private rights in water. In the American West, nineteenth-century judges developed the prior appropriation doctrine, which recognized rights to use water from rivers and streams. As its name implies, the doctrine favoured early diverters or appropriators over those who came later. This "first-m-time, first-in-right" principle divided surface water users into seniors and juniors, depending on the date when each person began to use the water. During times of shortage, the juniors were cut off while the seniors continued to receive their entire appropriation. Appropriators could divert limitless quantities of water so long as they used it for a beneficial purpose, most often for crop irrigation. The prior appropriation doctrine encouraged economic development by creating an incentive to divert water, but its environmental impact has been catastrophic-diversions have completely dried up many rivers. The doctrine left no room for competing values, such as the ecological, cultural, or spiritual importance of a free-flowing river to fish, animals, plants, and human beings. Although appropriators had legal rights to divert surface water from western rivers, they did not own the water itself. Instead, their property right was one of use, otherwise known as a usufructory right. The water remained a public resource owned by the state, which allowed individuals to use the water subject to government rules and regulations. The prior appropriation doctrine created very valuable private rights to water. But Mother Nature is fickle. In the West, river flows fluctuate, depending on the amount of snowfall during the winter and the rate of runoff during the spring snowmelt. Water is usually plentiful during the spring, when the farmers do not need it, and scarce during the summer, when they need it most. This is where the federal government came in.

Congress passed the Reclamation Act in 1902, creating the U.S. Bureau of Reclamation (BOR) and authorizing it to build large-scale dams and irrigation projects. The BOR literally changed the face of the West, from Hoover Dam to Grand Coulee Dam and from California's Central Valley Project to Arizona's Central Arizona Project. By the time the frenzy of dam building came to an end in the 1960s, most of the great rivers in the American West had been transformed into quiet millponds-storage reservoirs that served the needs of western farmers. None of these projects would have been undertaken by the private sector because they made absolutely no economic sense. But the federal government was less interested in cost-benefit ratios than in encouraging the development of agricultural communities throughout the West. Thus it built immense subsidies into the reclamation program.

In theory, the federal government would merely encourage the establishment of small family farms in the West by fronting the construction costs for BOR projects, which beneficiaries would repay once they began to use the water. The reality was something altogether different, and the payback provisions never operated to recapture the construction costs.

Despite the provision in the Reclamation Act that limited the size of individual farms to 160 acres, artifice and outright fraud enabled farmers to circumvent the maximum acreage restrictions and to amass agricultural empires of thousands of irrigated acres. Farmers' repayment obligations were based on their ability to pay, not on the actual costs of the project. Farmers' payments did not begin for 10 years, stretched out over 50 years, and were interest free.

To make the projects even more alluring to prospective farmers and irrigation districts, the BOR wrote off a substantial percentage of its construction costs as nonreimbursable federal benefits, such as flood control. In return for its huge investment in water infrastructure, the federal government could have insisted that it had rights to the water held in the reservoirs, but it did not. Instead, water developed by the BOR was doled out by contract to farmers and irrigation districts. Thanks to these federal dams, farmers and irrigation districts have extraordinary rights to a year round supply of water. The Bush

Administration recently decided to renew more than 200 contracts involving the Central Valley Project in California, the largest water-supply project in the country. By extending the contracts for 50 years and by committing a large volume of water to the project, the Administration has virtually guaranteed that Central Valley farmers will play a major role in California's water markets.

Farms, mines, and cities also have (largely unregulated) rights to groundwater, a situation that has caused horrible environmental problems. Groundwater pumpers do not own the water; instead, they have a right to use groundwater subject to government rules and regulations. A few states, such as Arizona, have enacted progressive legislation to curb unsustainable groundwater withdrawals, but reform does not come easily. Farms, mining companies, and cities have spent billions of dollars on wells. Entire communities and local economies have come to rely on groundwater. Turning the spigot off at this late date is not easy politically or legally. Legislators are reluctant to alienate powerful economic interests. Judges are loathe to disrupt the settled expectations of custom. Consequently, the groundwater table continues to drop.

In a handful of western states, the prior appropriation doctrine applies to groundwater. These states curb excessive groundwater use by granting legal rights to the first pumpers as against later arrivals. This doctrine avoids many of the environmental problems associated with unrestricted pumping because it coordinates ground and surface water use and avoids the tragedy of the commons. Yet, it does so by creating stronger legal rights. Though the rights remain ones of use rather than of ownership, they are enforceable legal rights.

In short, if we are to reallocate water from existing, low-value uses to newer, higher value uses, we must recognize that water rights in the United States are already privatized. Water rights in the American West belong mostly to individual farmers or irrigation districts, not to federal, state, or local governments.

As the population in the West mushrooms, local governments are scrambling in search of new water supplies. They face an uncomfortable reality: little water is available. Agricultural interests have a stranglehold on water in the West. Three points temper this ownership issue. First, most states have charged their water departments with the authority to consider the "public interest" when a new appropriator requests a right to make a diversion. In some states, this has helped protect the environment from the complete depletion of river flows. Second, the public trust doctrine which originally protected the public's interest in accessing navigable waterways for commerce, navigation, and fishing, has been expanded to include environmental protection and recreational access. Finally, federal environmental legislation, from the Endangered Species Act112 to the Clean Water Act, has mandated changes in state water rights. A notable example was the furor over the Klamath Basin in 2001. When a severe drought hit in 2001, the Secretary of the Interior decided that the Endangered Species Act obligated her to provide sufficient water for salmon to migrate downstream. The farmers received less water.

This action raised the Takings Clause issue, and the courts have reached different results. Most courts have decided that BOR contractors have no claim for Fifth Amendment takings compensation when they receive less water if it is necessary for other environmental needs. In one case, however, a court held that a taking had occurred.

During a drought in the early 1990s, the California State Water Resources Control Board decided to reduce deliveries of water to irrigation districts in order to protect endangered species. Those districts sued, claiming that the federal government had taken their contractual water right by imposing restrictions under the Endangered Species Act. In a poorly-reasoned opinion, the U.S. Court of Federal Claims agreed, and subsequently awarded the districts $26 million as compensation. In December 2004, the Bush Administration decided to settle the case for $16.7 million rather than to appeal the court's ruling.

Some may object to the idea that farmers who receive highly subsidized water through BOR projects can then turn around and sell it at 10, 20, or even 50 times what they paid for it. They declare this windfall unjust. Perhaps. But what is the alternative? If existing water rights holders cannot reap a profit on their water rights, those holders will not sell.

The situation is that simple. Given a choice between making a few farmers rich off BOR contracts or continuing the practice of using huge amounts of water to grow cotton, I think the choice is easy. Moreover, the farmers would be selling very valuable water rights-often several generations old-for use by cities and for the betterment of the environment. The measure of value ought not be what the farmers originally paid for the water or even what they currently pay.

The measure should be the value of the water in the hands of others. By that measure, the farmers are giving up something incredibly valuable, what Barlow and Clarke call "blue gold." Others may object that transferring water from farms to cities and suburbs encourages urban sprawl. Water marketing, some may argue, provides liquid nourishment to fuel growth. In situations where limited water supplies might constrain development, water marketing removes this obstacle.

Developers replace the open spaces of pasture land, cotton fields, and rice paddies with red-tile-roof subdivisions, Circle Ks, Walgreens, and the occasional Wàl-Mart or Home Depot. Preserving open spaces, including farmland, is a critical national priority. I'm no fan of mindless sprawl and surely no shill for developers. The problem with our existing water law is that we get both wasteful irrigation and mind-numbing sprawl.

We have not made a choice between one or the other. We have supplied water to serve both. That is why it is critical that we require new development to purchase and retire existing water rights: it breaks the relentless cycle of overuse and moves us toward sustainable water use.

Addressing the problem of water scarcity will require action on a number of fronts. We must encourage the reuse of municipal effluent, explore the technological boundaries of water desalination, impose appropriate conservation requirements, and raise water rates. We must also recognize the reality of private rights in water and embrace water marketing as a critical tool to reallocate water use. In this process, the government must play a critical role in overseeing water transfers to protect the interests of third parties and the environment.

6

Managing Water Diversion

Water is increasingly a major political issue as scarcity of the resource grips several societies, especially in developing countries where agriculture is responsible for about 80 percent of water consumptive use. However, western nations are not immune to water tensions, for instance in Greece, Spain, Italy, or the western United States, where the available water is being exploited to the limit: the Colorado no longer reaches the sea, and growing debates are emerging as to whether water should be allocated to thirsty cities or to agriculture; whether public funds should be invested again to increase the resource or demand management implemented; and whether water could be imported from far away.

These questions are increasingly relevant: consumption patterns of water in the western US are clearly not sustainable. Given the technology available today, massive water transfers could only come from Canada. There have been several mooted projects in this vein, mostly in the 1960s and 1970s. Should Canadians worry about water exports to the United Sates, especially in the frame of NAFTA? Or is climate change going to be the decisive factor in the debate?

The Advent of Water Addiction

Water is a key ingredient in the fabric of the western American society, as has been well studied by Donald Worster. The west is not completely waterscarce, for several mighty rivers flow in the region, mainly fed by snow and glaciers from the Rockies; but it definitely is a semi-arid region, compelling all societies living there before the industrial revolution to adapt to water scarcity. Early 20th century American society, empowered by the industrial age, decided to harness rivers and aquifers. Technology enabled American society to eliminate the water scarcity burden and developed the illusion that technology would always bring about a solution to growing water needs. "What nature does not yield freely, humanity should refashion to better suit human needs.... Nature has no greater purpose than to serve Man, and Man has no greater purpose than to work the land and take his place in the productive cycle."

Wendy Nelson Espeland also clearly depicted the representation that developed at the time that all water flowing unused to the ocean was a wasted resource, and that rivers needed to be "tamed" and "harnessed" to be put to use. President Franklin Roosevelt declared in 1935, when inaugurating Boulder Dam, that "the mighty waters of the Colorado were running unused to the sea. Today we translate them into a great national possession." This idea is still very much alive today. "Much of the contemporary culture of the American Southwest is based on denying its desertness.... In a subdivision being built to the south [of Las Vegas], Paseo Verde Parkway and Val Verde Road intersect in Green Valley Ranch. The concept of green, like sod lawns, was an imported fantasy."

Technology can make up for water scarcity and water must be used so as to develop western resources: "The rural Western ethic is that all wealth comes out of the ground, either as grass growing or as minerals being mined.... So the fact that today's reclamation projects-such as Garrison, CUP, Animas-La Plata-cost a few million dollars for each farmer they put on the land, doesn't cause their proponents to blink. That, they say, is the price society pays for creating the stuff of wealth. Without it and the other industries based on earth, there is nothing."

Early Massive Water Transfer Projects

Massive water transfers were first built in the eastern United States in 1847 with the Croton aqueduct for New York City and the 1900 Chicago diversion. As early as 1906, water transfers began being built in the west, with a 46 km-long canal from the St. Mary River in Montana to the Milk River, triggering a severe dispute with Canada that forced the United States to negotiate the 1909 boundary waters treaty. Other projects soon followed: the Los Angeles aqueduct in 1913, the Hetch Hetchy water supply (1934), the Colorado River aqueduct (1941), the Ail-American Canal (1942), and the California central valley project (1951). Diversions were initially designed for municipal water supply, but irrigation soon became their main rationale, as the sector benefited from generous subsidies from the federal government through the Bureau of Reclamation (created in 1902), its agency dedicated to the development of agriculture in the west.

When exploited resources began to show signs of exhaustion, engineering firms started to design huge transfer projects from Canada. But the main trigger of the idea that water resources were exhaustible and not adequate to sustain limitless development in the western United States struck in 1963 when the supreme court compelled California to stop excessive Colorado water pumping. Colorado River water had been apportioned between upper-basin and lower-basin states with the Colorado compact of 1922, but Arizona felt California tried to prevent its own use of the water. With the 1963 ruling, California's share was established at 4,4 million acrefeet (af), by far the largest share in the 15 million af available, but a figure that set a limit to what California could bank on. The very idea that the water resource was all of a sudden set at a strict limit sent shockwaves throughout California that triggered a frenzy of plans designed to increase supply through continental water diversion projects from Canada.

These projects, like the earlier water diversion schemes, involved diverting large volumes of water over long distances, but their sheer size placed them on a very different scale, both regarding their length (often greater than several thousand km) and their cost (several dozens of billions of dollars). They were designed, from 1959 to 1985, in the United States and sometimes in Canada, on a magnitude no project had ever reached before. These projects involved the transfers of the Mackenzie River, for instance, all the way through the flooding of the Rockies Valley to the Mississippi and Colorado Rivers, or the diversion of the Great Lakes, or even the damming of James Bay so as to convert it into a freshwater lake and its subsequent transfer to the American west. Costs involved reached hundreds of billion dollars and environmental impacts were not studied at the time.

Massive Continental Water Diversion Projects Never Implemented

Although these continental water diversion projects received considerable media and state government attention in the west in the 19605, none went beyond the feasibility study level and none managed to get an official approval.

Cost

Water is heavy. When transported by aqueducts over long distances, it requires a lot of energy merely to be moved and pumped over obstacles. The Grand Canal, designed in 1959 by Canadian engineer Tom Kieras would have required at least 6 nuclear plants merely to pump the water from James Bay to Georgian Bay in Lake Huron. It is therefore expensive to operate these infrastructures. Moreover, the capital requirement is huge, as attested by their estimated construction costs at the time of the project design. Governments were more than reluctant to invest such large amounts of money for water exports, especially when the profitability of such projects was far from certain.

Estimations of the cost of water transported or produced by different means show that desalination is becoming much cheaper than it used to be: about $0.6 to $1.2 per cubic meter, as compared to $0.55 to $1.35 for plastic container transportation; $1.25 to $1.5 with water-carrying ships; $0.6 to $0.85 with iceberg transportation; $0.07 to $1.8 for water recycling; and $0.8 to $3 for transfer canals of about 500 km. It is true these means do not all serve the same purposes, as desalination plants are usually located by the sea and serve urban demand. Sea water desalting technology, in particular, improved so fast between 1985 and the present that operation costs have been reduced dramatically and divided by five, from $2.5 to $0.55 per cubic meter.

Desalting is now a very affordable water-producing technology for urban and industrial consumers, enabling them to tap into an inexhaustible source. However, the water it produces cannot be considered for irrigation purposes, given its cost, its distance from the interior, and the large volumes irrigation demands. But these figures are useful inasmuch as they show that water transfer by canals is not as inexpensive as many advocates insist it is: it is not certain that farmers could afford their expensive water volumes without massive subsidies.

Stabilizing Demand in the United States

The main reason that the giant diversion schemes never came to fruition is that demand was not really present in the United States for Canadian water. Several factors explain why these projects lost their allure for both governments and society. The budget crisis that began in the early 19705 in the United States precluded public money for such large endeavours, at a time when public opinion gradually turned against large scale engineering projects that tampered with the environment: 1977 was the "hit list" year when President Jimmy Carter discarded several water development projects in the west because of their extremely low benefit/cost ratio. The economics of continental diversion projects have so far worked against them and will do so for several more years.

It was all the more difficult to argue for these continental diversion schemes as water use, contrary to what had been expected in the 1960s, showed a definite trend toward stabilization, especially in agriculture. More cost-recovery-oriented water pricing; competition from other regions, mainly Asia; and cost incentives that lure American producers to Mexico are among the factors that explain why water use in agriculture remained roughly stable between 1975 and 2000 throughout the country. If the federal government does agree, during the Doha round of trade negotiations, to reduce agricultural subsidies, water prices for farmers could increase markedly, resulting in financial incentives to consume less. Alternatively, water demand for irrigation could decrease as the result of farmers going out of business. As a whole, water withdrawals in the US have increased slowly, much more slowly now than its population, and could even begin a downward trend should competition from foreign fruit and vegetable producers work against local farmers.

In addition, although there still is room for improvement, water use per person is showing signs of stabilization, probably thanks to both tariff and education policies. The evolution of water use in the United States, from 1970 to 2000, shows that although the population grew by 38.6 percent, water use increased by only 10.7 percent. Urban consumers began to value water conservation with the development of urban water use reduction programs in most large cities. Industrial use actually decreased by 55 percent, whereas irrigation grew by only 6.1 percent.

Therefore, although water is still used at an unsustainable rate in the western United States, importing water from Canada is not as urgent as appeared to be to some politicians a few years ago, such as former senator Paul Simon. A stabilizing trend in water withdrawals and the availability of cheaper sources with desalting plants led public planners to forget about massive water transfers from Canada. When asked, even the Western States Governors Association estimated continental diversion projects were unlikely to ever be built."

Opposition from Within the United States

Moreover, there also was opposition from within. Governmental archives from the early 1980s attest to the western United States lobbying for the diversion of Great Lakes water to quench its thirst. The International Joint Commission, created to prevent and resolve disputes between the United States and Canada under the 1909 boundary waters treaty, explicitly warned against water diversions from the Great Lakes basin in its 2000 "Final report on protection of the waters of the Great Lakes."

Great Lakes states wanted to resist these projects, both for environmental and political reasons: why would the Great Lakes states give California added value at a time when so many firms were leaving the area and moving to the West Coast? The Council of Great Lakes Governors (CGLG), created in 1983, is a partnership of the governors of the eight Great Lakes states and the two Canadian provinces of Ontario and Québec. The Great Lakes charter stemmed from their growing concern that Great Lakes water could be diverted to water-scarce regions of the United States.

The charter, signed in 1985 by the CGLG members, created a notice and consultation process for Great Lakes diversions. The signatories agreed that no Great Lakes state or province would proceed with any new or increased diversion or consumptive use of Great Lakes water over five million gallons per day without notifying, consulting, and seeking the consent of all affected Great Lakes states and provinces.

After an Ontario firm, Nova Group, tried to tap water in 1998 from the Great Lakes for export purposes, triggering an uproar of protests from both the Canadian and the American sides of the lakes, the CGLG decided to further strengthen the water export ban. This led to the Great Lakes charter annex, signed in June 2001. The annex outlines a series of principles for reviewing water withdrawals from the Great Lakes basin that is grounded in protecting, conserving, restoring, and improving the Great Lakes ecosystem. In December 2005, the annex 2001 implementing agreement was signed, with specific guidelines for the enforcement of the principles. It will also form a legal basis for the ban when the Great Lakes-St. Lawrence River basin water resources compact is ratified by the US congress, making it legally binding for American Great Lakes states. With this agreement, water diversions are only permitted over short distances from the Great Lakes, must ensure that water is returned to the Great Lakes basin, and are only permitted if there is no other solution to satisfy the water demand.

Massive Transfers

Canadians became very fearful in the late 19905 about potential American water transfer schemes, leading to calls for the federal government to enact a ban on water exports. Wisely, Ottawa refused to enact any law that would implicitly recognize water as a good, but rather put forward bill C-6, voted in December 2002, which forbids interbasin transfers. Under NAFTA, this was a clever move inasmuch as it leaves water outside the realm of commerce and applies to both Canada and the United States. But Canadians were ignoring the fact that large-scale water transfers already exist throughout the world, and especially in Canada. Several massive transfer schemes have already taken place: in British Columbia (the Kemano transfer, 1954); in Manitoba (the Churchill derivation, 1976); in Ontario (the Long Lake and Okogi River transfers, 1939 and 1943); in Labrador (the Churchill Falls transfers, 1971); and in Québec (the Caniapiscau and Eastmain transfers, 1985).

There are major differences from American water diversion projects: Canadian transfer schemes were for hydroelectric development purposes, and not for urban consumption or irrigation. They are usually massive and over short distances, less than 200 km, whereas American water diversions often run for more than 500 km. There are currently plans to divert the Peace River in northern Alberta to meet growing agricultural and urban water needs, but that project is far from accepted. In addition, HydroQuebec is currently developing other water diversion projects for the Betsiamites, Eastmain, and Manouane rivers development.

There is controversy over the extent of the negative impact of these diversions on the environment, since the concept of minimum ecological flow remains controversial among biologists. Water does remain in Canada for each of these diversions, but this is not the point: Canadians collectively ignore the fact that their daily comfort and economic activity rest, to a certain extent, on major river diversions. It is therefore difficult to defend the argument that water transfers to the United States would weaken the environment, if one refuses to consider the phasing out of transfers within Canada.

At the end of 2005, with the signing of the annex 2001 implementing agreement, it was safe to say continental water diversions were definitely outdated as a water management concept. Demand was stabilizing in the western United States; public opinion had developed a strong distaste for the idea; governments were more than reluctant to invest the billions of dollars each scheme implied; diversion were by far not the most cost-effective way to provide for additional water volumes; and conservation and recycling were promising alternatives to the long-used supply management approach.

Toward The Return of Massive Continental Water Diversion Projects

Climate change could shake this status quo. As years go by, and as scientists as well as local governments accumulate impressive data attesting to a definite trend toward climate change, the developing weather shift could prove more challenging than the long-advocated water exports to the United States of the 19605 and 19705. The possibility that climate change could affect Canada's water quality and quantity is serious enough that Environment Canada is spending several million dollars on scientific studies to try and assess the risk. Competing uses-for urban areas, agriculture, and oil production, for instance-in regions such as Alberta where water could become scarcer, could lead to conflicts between groups that would see authorities intervene with unpopular measures such as water pricing or regulation.

Present forecasts, so far proven by observed data, show increased global precipitation, but this extra amount of water would be more than offset by increased evapotranspiration (Et) because of higher temperatures. In the already dry Okanagan Valley in British Columbia, water demand for orchards could therefore be multiplied by 12 in the summertime because of increased Et. In addition, a change in the pattern of how precipitation falls throughout the year, coupled with changes in temperature, could very much alter its availability, resulting in seasonal but dramatic supply reduction, as happened in Québec in August 2002.

In the Prairies, already less snow falls in the winter, and it melts earlier in the spring, allowing for less water in rivers in late spring and summer. Along the eastern slopes of the Canadian Rocky Mountains, glacier cover has decreased rapidly in recent years, and total cover is now approaching the lowest experienced in the past 10,000 years. Rivers in the Prairies are mainly fed in the summer by melting glaciers. As the glacial cover has decreased, so have the downstream flow volumes. This finding appears to contradict projections of the intergovernmental panel on climate change (IPCC) that warmer temperatures will cause glacial contributions to downstream flow regimes to increase in the short term. Historical stream flow data indicate that this increased flow phase has already passed, and that the basins have entered a potentially long-term trend of declining flows. The continuation of this trend would exacerbate the water shortages that are already apparent across many areas of Alberta and Saskatchewan owing to drought. Climate change is of course not merely affecting Canada, but the United States as well. Similar conclusions have already been reached with regard to the potential impact of the Rockies' disappearing snow and glacial cover on the western United States. Planning for the impact of climate change remains difficult, because building accurate models for various regions-ones that give a true picture of future available runoff-remains a daunting task for programmers. What we now have are educated guesses sustained by general models and data trends. But these trends do show that several regions in the United States, especially in the dry west, could be faced with critical water shortages, given present demand.

This might prove to be a scenario in which the United States could become interested again in Canadian water: the goal would be to prevent a social crisis triggered by conflicts between agriculture and cities for dwindling water resources. The equation for the American federal government would be the perceived financial, social, and political costs of finding new water resources, as opposed to arbitrating against agricultural water needs. No one can predict what choice officials would then make if this scenario came true. The following parameters would be significant:

- Agriculture is responsible for 80 percent of water consumption in the American west.
- A 15 percent reduction in irrigation use would satisfy the needs forecast for cities and industries."
- Agriculture does not currently pay the market price for its water, thanks to strong federal subsidies; tariff increases would result in more disciplined behaviour and encourage conservation.
- Western farmers have begun selling or transferring their water rights to cities, as illustrated by the October 2003 water deal between the Imperial Irrigation District and the Metropolitan Water District in California.

Massive water diversions are not new in North America: they were extensively used in both Canada and the United States, in the latter mainly to develop the west and to bring water to cities and fields so as to sustain an imported way of life, and in the former to develop hydropower projects.

Beginning in the 19605, western American states and engineering firms began developing continental water diversion projects because of fears that continued development would bring about severe water scarcity and social disruption. These doomsday scenarios did not materialize, due to reduced economic growthespecially in agriculture, as well as more efficient water use technology and the advent of water conservation, and the huge environmental and financial costs of these projects.

However, the very idea of massive water diversions is not completely dead. Tom Kierans, the designer of the Grand Canal, is still lobbying Ottawa and Washington. More challenging is the idea that ongoing climate change could dramatically affect agriculture in the American and Canadian west. Under these conditions, what answer would governments have? Whither agriculture? Would governments favour the more profitable and water-efficient farmers or bring back to the forefront the idea of transferring large amounts of water from humid parts of the continent to drier zones? The answer probably lies in the severity of the developing weather pattern change and in the adaptation capability of the agriculture sector.

Domestic Water Supply in Rural Areas: Implications for Policy

Lack of access to safe water is at the heart of the poverty trap, especially for women and children, who suffer in terms of illness, drudgery in collection of water, and lost opportunities because of the time that water collection consumes. In rural Africa, according to the World Bank, 40 million hours are spent, each year in collecting water for domestic use and half of Africa's population is without access to safe water (Black, 1998). Recent renewed focus on poverty alleviation has resulted in increased attention to the benefits of improved water accessibility. Poverty assessment research has consistently shown that improvement in water services is a critical element in designing and implementing effective strategies for poverty alleviation. In Tanzania, experience with respect to rural water improvement has, in general, been an unmitigated disaster.

Water provision facilities have invariably fallen into disuse or disrepair because the approach used to run them failed to ensure sustainability of services. As a result, a new water sector vision has emerged based on the demand responsive approach (DRA). The major principle of the vision regards water as an economic and social good to be managed at the lowest appropriate level. The DRA recognises the inherent capacity of communities in taking greater responsibility for identifying and solving their water supply problems.

Adoption of the DRA by most water and sanitation programmes has changed the approach to evaluating their success and failure. Traditionally, water and sanitation programmes have depended on detailed blueprints to provide the basis for control and predictability. Benefit-cost analysis has been a major economic tool in evaluating these projects. However, the DRA broadens the scope of evaluation and blueprints cannot be drawn up since decisions are made jointly with communities, and problem solving is based on those partnerships (Narayan, 1995). Participatory evaluation becomes an essential tool in assisting attaining stated objectives and is an integral part of overall programme monitoring and evaluation activities. Participatory evaluation requires the utilisation of different tools and approaches in evaluating the sustainability of such projects.

This chapter is part of a follow-up participatory evaluation exercise of two community water supply programmes in Central Tanzania and involves presentation of the results on communities' willingness to pay (WTP) for improving the availability and quality of water services. WaterAid, a non-governmental organisation (NGO) based in the United Kingdom and the Lutheran World Federation through the Tanganyika Christian Refugee Service (TCRS), provide substantial support and expertise for such programmes. The former operates in the Dodoma Region while the latter operates in the Singida Region. The two regions constitute the semi-arid plateau of Central Tanzania.

It is anticipated that positive WTP indicates potential community ability to recover operation and management costs (Altaf et al., 1992). Thus WTP can be used to help ascertain the potential for fulfilling sustainability, at least from a financial viewpoint. A major objective of the study was to find out if communities studied were willing to pay an increased user fee or tariff for improving and expanding existing water services. If willing, how much (i.e. what percentage) could be added to the current user fee or tariff to achieve the required improvement? Based on the random utility hypothesis, a system of Multinomial Logit functions were developed and used to estimate individuals' willingness to pay and impact on revenue from increasing the water user fees or water tariff. The WTP was calculated by estimating the payments that would cause the respondent to be indifferent after a single unit increase in an independent variable. The estimated mean WTP and proportion of individuals who are able to pay were used to calculate the potential revenue that could be generated if the policy of increasing user fee or tariff was instituted.

In Tanzania, only 40% of the population has access to clean and safe water (MoW, 1997). Between 1965 and 1985, the Government of Tanzania (GoT) spent a substantial amount of money in developing rural-based water prefects to alleviate the water supply problem, but without notable success. Most of the investments ended in failure. Government, in collaboration with NGOs, is currently trying to revive many of these projects.

The government is providing technical assistance, while the NGOs supply initial capital requirements, overhead costs and supervision in their design and implementation. When the utility is operational, the beneficiaries (i.e. the villagers) either pay for the water from the utility or contribute an annual water fee, both of which are added to the village water fund. It is anticipated that the size of the village water fund will be sufficient to operate and expand the water utilities, based on market forces. To attain such sustainability-related goals, existence of efficient water markets is essential. Sufficient funds have to be generated to meet both overhead costs and investment requirements. Information on WTP for required improvements provides an indicator of the communities' potential capability of generating funds for such purposes.

Tanzania has sufficient surface and ground water to meet its present needs. Lack of capital and uneven availability of water across regions has limited Tanzania's capacity to utilise existing water resources to provide her population with clean and safe water. As population increases, conflicts in the allocation of water between energy, households, commercial livestock, and use in the agricultural sub-sector, increases. Weak water policies, inadequate community participation, and uncoordinated donor support has resulted in wastage or inefficiencies in the form of failed water development programmes.

After independence in 1960, donors supported large-and small-scale water development schemes. However, communities participated little in their planning and management, and also contributed little to both capital and operational/management costs. The policy of the government at independence was 'free, clean and safe water for all,' the objective being to provide clean and safe water to all villages in rural Tanzania by the year 2000 (MoW, 1997). As indicated above, this policy has now been abandoned in favour of a demand-driven water development programme. The new water policy enacted in 1997 (MoW, 1997), promotes the provision of efficient, affordable and sustainable water supplies and sanitation.

The emphasis is on community planning and management, private provision of goods, works and services relating to water supply, and public sector regulation, facilitation and environmental management. Water supply or water development programmes have to be demand-driven, and the community must be willing to participate in decentralised management and cost sharing. The Ministry of Water (MoW) has been restructured and its role has evolved from that of a water service provider to that of a regulator and a facilitator.

The new water policy underscores the importance of community participation and management in ongoing and new projects. In order to improve water supply services at the village level, the roles of MoW and other stakeholders can be summarised as follows (MoW, 1997):

- Village level. Small-scale water supply projects are to he operated and managed at the village level. Operation and management costs are to be met with funds raised within communities. The emphasis is on management by participation through formulation of village water committees that oversee and manage the utilities on behalf of community members. The communities agree upon the operational modes, with specific emphasis being paid to women's participation at all stages of water project development and management.

- District level. The water department at the district level is required to facilitate training of water managers and attendants/mechanics at the village level. The department is expected to maintain a pool of experienced technicians who will collaborate with village level mechanics in servicing and repairing established utilities. Another role of the water department at the district level is to make sure that necessary spare parts are available when needed.
- Regional level. The water department at the regional level has jurisdiction in providing guidance and making sure that government water policy and rules are adhered to. Apart from managing water supplies at the regional headquarter level, the department has to provide consulting services to districts and villages in terms of training, and facilitating availability and distribution of spare parts.
- National level. At the national level, the MoW has responsibility for financing and managing large-scale water supply programmes (especially in large cities), to train water professionals, and to finance maintenance units at the district and regional levels. Other responsibilities are to standardise capital equipment and tools used in water development, to help facilitate availability of spare parts, and to coordinate donor and NGOs activities to make sure they follow Tanzanian water policies and local government rules and regulations.

The general objective of the water policy is to propagate the new vision of community participation and management based on the DRA. Communities are expected to appreciate that water supply is no longer a free service provided by government, but rather government or other development agencies are available to facilitate and complement community efforts at meeting their own water supply needs. Thus, this vision is currently guiding the development of community-based water supply projects at the village level in the Dodoma and Singida Regions of Central Tanzania. Detailed descriptions and the rules pertaining to the programmes in each region can be found in Kaliba (2002) or the WaterAid website.

In both regions, financing for the water projects is thus divided among the beneficiaries, the government and donors. Villagers make cash contributions towards capital costs. They contribute time and labour, local materials and contribute to hospitality costs of visiting government staff. The initial amount of money contributed to a project by a village has been standardised and agreed by the major donors working in the regions. The cash contribution is not currently expendable but acts as a reserve fund or safety valve for the project. The community-based water fund has to be increased using profit margins generated from user fees (Dodoma Region) or tariffs (Singida Region) when the project is operational. The government provides professional staff, an annual cash contribution towards the programme, and provides some transport and most of the construction equipment. Donor (WaterAid in the Dodoma Region and TCRS in the Singida Region) funding pays for the purchase of locally procured materials, government staff fieldwork allowances, training courses and running costs. The donors also procure all imported materials, equipment and vehicles, and employ technical and management back-up staff. In the Dodoma Region the village water utilities have emphasised deep boreholes and engine-driven pumps, and charge a tariff (i.e. a specific amount per litre of water used). In contrast, in the Singida Region, much greater emphasis is placed on shallow wells (8-10 m deep) and a user fee (i.e. a specific amount per year per household), is charged. In the Dodoma Region, between 1991 and 1999, 857 projects were implemented under the WAMMA1 project funded by WaterAid. During this study period, 29% of the projects were not functioning. For those not working, 75% were waiting to be repaired, and 25% had permanent problems (e.g. missing engines, damaged pumps or dried up waterholes).

Useful indicators of progress arc thc growth in the populations served and village water funds voluntarily contributed by communities. The total population served increased from 0.27 million 1994 to 1.23 million in 1999. This is an increase of about 320% over a period of six years. During the same period, the contributions to the village water funds increased, in real Tsh terms, from about 0.24 million in 1994 to about 14.87 million in 1999. Since there is no evidence of a regional increase in purchasing power, and given the low income per capita in the region (BoT, 2000), this is a good indicator of the success of community-based water development projects in the region.

During that period, TCRS constructed 49 boreholes, 465 shallow wells, and three windmills. The number of villages with water committees increased from three in 1986 to 244 in 1995. In the same period, three villages did set up water committees without the help of TCRS. The number of villages with water funds increased from zero in 1986 to 229 in 1995. Although these numbers indicate the magnitude of time and money invested in water utility projects, they of course, do not provide by themselves any explicit evidence of the potential sustainability of the projects and/or progress towards achievement of sustainability goals.

To elicit WTP information from individuals, the contingentvaluation (CV) method is used. There are three major procedures used in CV surveys, namely conjoint analysis, dichotomous choice and the payment scale approaches. The conjoint analysis procedure asks respondents to rate rather than to price alternatives. It allows for measurement of consumer preference between items with multiple attributes (Baidu-Forson et al., 1997). In the dichotomous choice approach, respondents are asked whether they would vote to change the provision of some public good at a cost of $X to themselves. The respondents answer yes or no. For the respondent who answers yes, the experiment is repeated by varying X, until the respondent says no. After that, the respondent is asked to state the value he/she is willing to pay for the good or service if provided. For the respondent, the amount mentioned is considered the maximum WTP for that good or service. The payment scale procedure allows respondents to choose a value or price (cost) on a given scale measure. A range of values are presented to the respondents and they are asked to identify one value or price (cost) they are willing to pay or incur to purchase the good or service being evaluated. The respondent's choice from the given scale indicates the WTP for the stated good or service. Positive values indicate positive demand, and zero values indicate zero demand for the good or service.

In all cases, the surplus benefits are calculated using the parameters estimated from a model based on the type of distribution attached to WTP. In most cases, the models and type of distribution to use are dictated by the type of questions imposed and responses obtained. Other good references are those of Rosenberger and Walsh (1997) for ordinary least squares models; Kenkel and Norris (1995) for models with log-normal distribution; Steven et al. (1997) for Tobit models; and a classical work by Haneman (1984) for Probit and Logit models.

CV surveys entail three operations: designing a questionnaire, conducting a survey, and analysing the results. The validity of the estimates depends on how skilfully these operations have been done (Christe and Schwab, 1995). Donaldson et al. (1997) suggest that the use of a payment scale is more valid than the open-ended or closed-ended approach when presenting hypothetical bids. The limitation of the payment scale is constructing scaling benchmarks. This limitation can be reduced through using available market and nonmarket information to construct the scales, as suggested by Donaldson et al. (1997).

From the above review, it can be seen that there are different approaches to modelling WTP. Nevertheless, all procedures involve the respondent choosing one option from a range of other alternative services or goods based on their expectations. In order to develop models that are consistent with economic theory, the McFadden (1981) random utility hypothesis has become a standard approach to modelling WTP. As far as our study was concerned, the objective was to find out if communities are willing to pay an increased fee or tariff in order to get additional money for improvement or expansion of existing water utilities. If willing, how much (i.e. what percentage) should be added to the current user fee or tariff to achieve the required improvement? Field data were collected from 30 villages that participated in community water utility projects for a period of three years in Dodoma and Singida Regions (15 villages in each region) prior to the field surveys. Dodoma Region was divided into five clusters based on tribal distribution. From each cluster, three villages were randomly selected, differentiated according to accessibility. The results of past participatory rural appraisals (PRAs) conducted by WAMMA (some as early as the 1999) were reviewed and used in making adjustments to the questionnaire and in helping to develop pre-coded responses and socioeconomic indicators. In the Singida Region, due to transportation problems (part of the El-Nino legacy), only Singida Rural District was included. Rased on information gathered at the District and Regional Water Departments and the Regional Development Office, the district was divided into three agroecological zones. From each zone, five villages were identified to form a sample cluster. The Singida Regional Development Officer also had some PRA results that were used to fine-tune the questionnaire used in the study.

Two survey instruments were developed to gather the necessary information — the village checklist and the structured questionnaire. The village checklist solicited general information at the village level. This was done through meetings and group discussions attended by community members, government leaders, project managers, pump operators, extension agents, and water committee members. The major objective of the checklist was to get a general overview of the village projects based on group discussions. The discussion also helped to pre-test the structured questionnaire and in further adjusting the precoded responses. The questionnaire was administered to 225 individual respondents in each region. In each village, 15 respondents (at least five men and five women) were interviewed. The respondents were selected randomly from the village register and grouped based on age and social status of the individuals. Sample selection was stopped when there were enough sample respondents in each group. The two survey instruments and further information about the approach used in the study can be found in Kaliba (2002).

Regarding information on participation and management, several questions were asked. The questions related to the source of the project initiative, personal involvement in decisionmaking, labour and financial contributions, and the involvement of women in project design and implementation, emphasising participation in decision-making and informed choice. Questions were also asked regarding consumer satisfaction with the services and management of the water utility. The questions focused on: getting information on personal involvement in management; rating of the water managers' performance; knowledge of how the water management structure functioned; and the possibility of influencing change. To quantify respondent participation in project activities, and satisfaction in project performance, individual's responses were aggregated and used to calculate participation and satisfaction indexes as suggested by Sara and Katz (1998) and by the World Bank/UNDP (2000). Once again, further details on the approach used as far as this chapter is concerned can be found in Kaliba (2002).

On the WTP section, a respondent was asked if there was a need to improve the current water service. The answer was either 'yes' or 'no'. If yes, what kind of improvement was needed? The respondent was encouraged to identify only one improvement. After the respondent identified the type or choice of improvement required, she/he was asked to suggest an increase in tariff (for the Dodoma Region) or user fees (for the Singida Region) in order to cover the costs of the proposed improvement.

As explained above, the analysis was directed at identifying types of water services that need improvement and ascertaining the WTP for the improvement. In the Dodoma Region, the results concerning the water utility service can be summarized as follows: about 14% of respondents indicated that they were satisfied with the status quo (no improvement); and 81% suggested increasing pump capacity or buying a more powerful engine (thereby increasing the rate of water discharge). Thirty-three percent of respondents proposed increasing the number of water distribution points (water reticulation), and 22% proposed other improvements such as fencing, construction of new water tanks, construction of security houses, and construction of livestock watering points. The suggested tariff increase ranged from 5 to 100 Tsh/20 1. The existing tariffs averaged 20 Tsh, with a standard deviation of 9 Tsh (per 20 1 bucket of water) and on average amounted to a total of 895 Tsh/household/year.

In the Singida Region, 31% of the respondents were satisfied with the status quo, 80% wanted deeper wells that can produce more water, 29% wanted to increase the number of hand pumps in their village for water distribution purposes, and 10% indicated other types of improvement enumerated above. The suggested increase in a user fee per household per year ranged from 20 to 200 Tsh. The average user fee was 503 Tsh per year with a standard deviation of 170 Tsh. During the analysis, variables for Clusters 4 and 5 in the Dodoma Region and Cluster 3 in the Singida Region were not included in the models due to correlation problems. The likelihood ratio statistics that test for no effects was statistically significant at the 5% level in both the Dodoma and Singida Regions models.

The positive and statistically significant variables in the Dodoma Region were family size and satisfaction in the performance of project activities. These results imply a respondent residing in a relatively large family or who was satisfied in the project's performance were more likely to answer 'yes' to the question regarding the need for water service improvement. A large family implies the need for more frequent water collection trips. Improvements relating to reducing congestion at watering points (increased pump capacity or number of watering points) are likely to reduce time and effort expended in water collection. Satisfaction with reference to project performance implied both an increased demand and willingness to commit resources for improvement.

Negative and statistically significant variables were age, wealth and cash contributions. Older people and richer respondents were thus more likely to choose to maintain the status quo. In the study area, older people are less likely to be directly involved in water collection activities, while rich respondents could conceivably have access to other water sources (e.g. have their own wells) or delegate others to collect water for them. The negative sign on the cash contribution variable can possibly be attributed to 'contribution fatigue.' Respondents who contributed more during project initiation or development were more likely to say 'no' to the improvement question.

Finally, respondents in Clusters 4 and 5 were more likely to be predisposed to commit resources to improvements, because the general performance of the projects in these clusters was relatively poor (Kaliba, 2002). The underlying implications of the above results for water utility projects in the Dodoma Region are that projects whose performance creates a sense of satisfaction among consumers and a strategic water fund that is rising (so as not to create contribution fatigue) are critically important ingredients in generating community-related resources for water-related improvements. In the Singida Region, females were more willing to pay for improvement than male respondents, not surprising since they are primarily responsible for water fetching activities. The fact that this variable was not statistically significant in the Dodoma Region was unexpected. However, in the Dodoma Region water utility projects produced a more reliable water supply and there were generally fewer complaints about their performance (Kaliba, 2002). Another important and positively significant variable was the level of satisfaction concerning the performance of the project.

The Multinomial Logit model was used to model factors affecting choice of services: maintain the status quo, increase rates of water discharge (pump capacity or depth of water well or more powerful engines), increase the number of water distribution points/water reticulation, and other improvements, as explained above. In constructing the Multinomial Logit models, the parameters of status quo variables were set to zero. Again, the signs on the estimated parameters show the direction of the marginal effects on choice of improvement desired. The comparison is related to the deleted choice. As an example, for Dodoma Region, respondents who perceived that water service quality was high were more likely to say 'no' for an increase in water supply or the number of water distribution points but were more likely to say 'yes' for other improvements. The interaction between quality and proposed bids were important in making the choices or indicating the type of improvement desired. The opposite was true for the Singida Region where the interaction between quality and bids influenced the choices negatively. There, the important factor that significantly influenced the choice of improvements was individual bids. This may be related to the technology and poorer level of services in the Singida Region. Respondents may not be willing to spend more money on failing projects. Singida Region projects will have to increase performance in terms of quantity and quality of water supplied, in order to motivate community members to contribute more resources toward project development and expansion. This was particularly important in some villages where community participation in project initiation development was low (Kaliba, 2002).

Again, the reference is the status quo bid. The value of-0.145 in the quality variable in Cluster 1 of the Dodoma Region implies an increase in the quality of water by one unit decreases the percentage of respondents who want increase in water supply by 14.5%. About 64% of respondents in Dodoma and 59% of respondents in Singida voted for an increase in water supply through increased pump capacity or number of water distribution points (water reticulation). Most villages have only one watering point and it is clear that a lot of time and effort is required for water-collection activities.

Estimated Mean Willingness to Pay

Respondents who wanted to increase water supply in Dodoma Region were willing to pay 82 Tsh per 201 of water above the existing tariff. Respondents whose choice was improving water reticulation possibilities and other types of improvement were willing to pay 29 and 5 Tsh above the existing tariffs. In the Singida Region, the analogous amounts were 91, 68 and 80 Tsh per household per year, respectively. A joint t-test failed to reject a null hypothesis that there is no gender difference on willingness to pay in both regions at the 5% level of significance. In the study area, water fetching is not a man's responsibility. However, they care. Time wasted by women and children in the search for water has a negative impact on their other activities and therefore has a compounding effect on all members of the household.

Potential Revenue from Communities

Potential revenue that can be generated for project improvements was estimated using the estimated mean willingness to pay, proportion of respondents who chose the services, and reported average daily consumption of water per household. It can be seen that there is the potential for generating much more in the way of community-derived funds in the Dodoma Region compared with the Singida Region. For the Dodoma Region, the amount that could be generated is substantial by Tanzanian standards. In fact, the amount may be more than the budget allocated by the central government to regional development programmes in all sectors of the economy. However, a note of caution is in order.

Care will be needed to implement policies aimed at increasing revenue generation for water improvement purposes. Higher tariff or user fees may exacerbate inequitability of access to water services because poorer households will undoubtedly be more price-sensitive than richer households. Although there are potentially mechanisms to subsidise the poor and the elderly, such arrangements tend to alienate those affected and limit their willingness to participate and be involved in decision-making processes.

Policy Implications

Results for the WTP analysis and on proposed choices for improvement, imply that respondents were eager to improve the ease and efficiency of the water collection process. For most respondents in both regions this was explicitly articulated in the form of a desire to increase water supply and the number of water distribution points. Most villages have only one or two watering points, and it was clear that a great deal of time and effort is expended in water collection activities.

There is some potential for increasing project revenues to address such needs through increases in water tariff or water user fees. However, duc to the differences in the technologies to access water, it appears that substantially more funds can potentially be generated for such purposes in Dodoma than in Singida. The water technology being used in the Singida Region is apparently not solving the problem of the dry season water scarcity, let alone the issue of water quality. A new approach and vision is needed to provide a long-lasting solution to the problem in that region. Dependency on shallow wells seems to have less potential for contributing to financial sustainability of water ulility projects in the Singida Region than the deep boreholes in the Dodoma Region.

Whatever happens, in both regions, if increases in water utility levies are planned in order to improve water services, two issues will need to he carefully considered. One is to take into account the views of women especially with respect to the location of new watering points or wells. The other is to ensure that any adjustment in water levies does not create undue hardships for the community or unduly inhibit access by the more disadvantage families/ individuals in the communities (i.e. the poor or elderly).

In conclusion, project sustainability, at least from a financial viewpoint, will be largely determined by the degree to which it continues to deliver its intended benefits over a long period of time. As far as Central Tanzania is concerned, there is some potential for developing sustainable water supply programmes based, to a large extent, on a community cost recovery mechanism.

In villages where there has been strong satisfaction with respect to water project performance, this chapter has demonstrated that individuals are willing to contribute more resources for improvement. It is important to note that, in these villages, community members are highly motivated and participate strongly in the daily management of the projects. An increase in community participation and changing of the water provision technology for the Singida Region is important for future project expansion and new project development.

Role of Virtual Water in Solving Water Scarcity

Water, as a source of food security, plays an essential role in ensuring sustainable food resources for a growing population. However, water scarcity is a common problem in many countries and regions. Because of the uneven distribution of water resources and population densities worldwide, water demand already exceeds water supply in nearly 80 countries that support more than 40% of the world's population (Bennett 2000). Water scarcity has become an increasing constraint to their economic development, particularly food production-the biggest water user (Hong et al. 2002).

Nonrenewable fossil water has been exploited to relieve the immediate pressure of water stress; this use has caused the depletion of resources and obstruction of future economic development and food security. To combat water scarcity and increasing food production, many countries have made efforts to expand irrigation. Despite a tremendous effort, the coverage ratio of irrigation in most of these countries remains low, simply because of their scarce water resources. The growth of food production overall has lagged and has been unstable (FAO 2001). In contrast to food production, the growth of population and food demand has been rapid in China. Meeting the water-food challenge is difficult in China because of the increasing population and water demand. In order to relieve water stress, many investment projects in China are planned and have been carried out, in order to focus on changing the uneven distribution of water resources, and to fill the gap between water supply and demand.

The role of virtual water through food trade as a solution to water scarcity is not recognised in China. Indeed, most researchers do not associate the international food trade with food security. However, they now advocate regional food trading as a solution to water stress in Northern China. These opinions are consistent with Chinese government policies of food security. The considerations of the Chinese government for food trade include domestic food security and social stability, the impact on rural employment when grain production remains the major provider of employment in rural areas, and the pressure of rural-urban migration and the social economic impacts on cities (Hong 2002).

Water Deficit and Demand

Water Scarcity in Agricultural Production

The increasing demand for food needs two basic natural resources for its production, arable land and fresh water. Future food production depends on the availability of adequate and sustainable water resources. Agriculture is responsible for about 70% of all freshwater withdrawn, and it is expected that more water will be used for irrigation as world food production increases.

An FAO study of 93 developing countries indicates that some water-scarce nations are already withdrawing water supplies faster than they can be renewed (FAO 2004). Over the years, China has maintained a population increase of 1.4% and an increase in economic development of 9% per year through using a great deal of water with low efficiency. However, agricultural irrigation is constrained by water scarcity and pollution, with 70% of food output produced on irrigated land, and 73% of the total available water used for irrigation. Furthermore, agriculture is in fierce competition with industry, municipal, living and environment water uses.

Actually, in China, agriculture competition for water is below industrial and domestic demands (Hong 1999) because the direct economic output of agricultural production is rather low. In China, every 10^4 m^3 of water feeds 4.46 persons.

North China, Inner Mongolia, Hebei, Beijing, Tianjin, Shanxi, Henan and Shandong, seven provinces, use about 168.5×10^9 m^3, which only can feed 75×10^6 persons, i.e. 23% of the total population. The demand for food by the population is met by food produced in Southern China, where farmland is scarce although water resources are relatively sufficient. Present resources meet the demand for food by the large population. Indeed, the requirement for freshwater resources is four times higher than what is normally available. The required capacity for freshwater resources is the lowest of all the resources estimated (Qu 1992).

At the national level, it is estimated that there may be enough water available, but some regions will face serious water shortages, with inefficient use and serious pollution. In northern areas, water scarcity and water pollution are serious while floods and water pollution are the main problems in southern areas. 95% of available water resources in the north have been exploited, while 6% of water resources have been used in the south. Water consumption in agriculture increased in North China where land is not a limiting factor for food production. Because water use efficiency is on average 0.4-0.5, irrigation water in the north is 7.5×10^3-12×10^3 m^3/ha, or 2-5 times higher than actual crop water demand (Liu 2000). The management of water use is expected to improve for avoiding water waste and enhancing irrigation capacity.

Water scarcity in the north has resulted in higher opportunity cost than that in the south. The huge investment in the South-North Water Transfer Project also supports this viewpoint. For water-scarce China, water use efficiency is much lower than that of developed countries. With the rapid population growth, water stress is becoming a formidable challenge. Meanwhile, when water resource load capacity reaches the limit, direct cost of enlargement of food production will quickly increase and result in native food price rises. The most plausible result is that the cost of water for agriculture becomes much higher because of competition between sectors.

Future Water Demand of Agriculture

Agricultural water consumption is the principal source of total water consumption, and is increasing with growing food demand. Irrigated area has increased, but total irrigation water quantity and withdrawal do not balance. This is due to irrigation efficiency enhancement and new irrigation technology improvements. Total water consumption has increased with unchanged available agricultural water quantity. The increasing water demands of economic development and domestic needs make additional water for agriculture more difficult to find, especially in North China. The large acreage of crop production in this area indicates a large irrigation requirement.

Because different crops have different irrigation water demands and different arable land acreages in North China (Qian et al. 2001), total irrigation water demand is $48.55 \times 10^9 m^3$; 5.00×10^9 m^3 for rice, 26.53×10^9 m^3 for wheat and 17.02×10^9 m^3 for maize. Actual irrigation efficiency in agriculture is quite low (0.4-0.5), thus, actual water demand required to support production in North China is about 97.1×10^9 tonnes. The ratio for agriculture water use accounts for about 57.63% of total water resources in North China, which irrigated only 53.12% of its total ploughed acreage. The remainder of the ploughed acreage irrigation demand, as well as other sectors' demands, intensifies water stress, influencing economic development and agricultural production. A water transfer project was selected to alleviate water pressure in North China, but in the view of the economic cost, virtual water trade through food trade may be a better solution for water deficits.

Food Demand

Population Growth

The population in China is about 1.3 billion and has been increasing. Expansion of population influences every aspect of social development. Insufficient natural (fresh water, land, forest, etc.) and productive resources (e.g. food) originate from population growth. Finite resources cannot support the huge demand of the growing population. Resources allocation ratios are very small. These problems interlace and restrict economic development and social advancement. At the same time, potential risks are concealed in the problems. The population control policy, which confines the number of children in a family, cannot restrict overall growth in the short term because the growth is based on an existing huge population. It is predicted that population growth will be negative in 2050 (Kang 1998).

The most important problem is that China must have enough food for the population, whether from food production or imports. Freshwater resource scarcity and uneven distribution strain food production, resulting in an increasing portion of domestic food supply coming from imports. Such imports, in essence, have substituted for the water that would otherwise be needed for producing the food locally.

Studies of virtual water and food security in Mediterranean and Africa countries have shown that water scarcity in these countries resulted in food insecurity. China has the same food issues, and uneven distribution of water and land resources has resulted in food supply scarcity. Furthermore, there is rising competition between water needed for agriculture and water for domestic use in burgeoning cities and ever-expanding industrial developments. Consequently, today the world must act wisely to conserve, preserve and improve its water supplies (FAO 2004).

Food Demand

Food demand is a function of the size of population and consumption per capita. Rapid population growth will inevitably drive up future food demand, while per capita food consumption is likely to increase as income rises, adding an extra demand for food (Hong et al. 2002). The growth in population and foodstuff demand per capita has jointly promoted the growth of total foodstuff demand in China (Kang 1998). Foodstuff demand requires that food supply increases concomitantly. However, expansion of food production is limited by farmland and water resources; thus, the pathways of food supply enlargement are expected to change. Increasing output and food import is required to solve food scarcity, together with water resources, to meet the food demand of the population. Water resources and food trade are complex issues for society.

Crop Production and Food Imports

Crop Production and Food consumption

Cereal production in China provides the basic energy requirement directly as food and indirectly as feed for animals that are used for human consumption. The increase of grain production, including rice, wheat and maize, has been rapid since 1980. The trend in grain production up to the end of the 1990s was smooth, with only small increases from 1991 to 2000 because land productivity is approaching the limit, and no new productive technology has been created to promote increasing food production. The production of cereals, cotton, oil and sugar was either unchanged or declined, while the production of fisheries and livestock products has rapidly increased.

Before 1997 most cereal production was consumed domestically, typically rice, since then, the diet has changed and more non-cereal products are being consumed. With increasing income, meat, egg and milk demand has increased while cereal consumption has decreased.

In contrast to decreasing direct cereal consumption, meat consumption has resulted in indirect cereal consumption by livestock. At present, according to the current GDP per capita, China is positioned by the World Bank in the low-middle income category. However, because of rapid economic development, per capita dietary structure indicates that life quality change is one of the most important variables that influences consumption.

Food Trade in the Past Two Decades

Food imports in China have fluctuated, especially wheat imports. This import demand is decided by domestic production capacity, and is a way to ensure food supply. Production technologies, especially water saving technology and multiple crop index improvement, have promoted the productivity of wheat in China. However, the increase in food production is still limited by water and land resources.

Although irrigation efficiency has generally improved, insufficient water resources are allocated to agriculture. Food products imported into China from 1980 to 2000 were mainly in the form of wheat and maize. Net import volume has decreased gradually for wheat from 11.59×10^6 tonnes in 1995 to 0.88×10^6 tonnes in 2000, and for maize from 5.18×10^6 tonnes in 1995 to zero in 2000, because domestic supply has increased. However, in contrast to declining imports, the volume of export products has increased from 1980 to 2000 due to increasing production because of improved agricultural technologies. Growing export products include rice, cotton, pork and fishery products, in which rice export rose from 0.05×10^6 tonnes in 1995 to 3.75×10^6 tonnes in 1998. Pork exports declined from 3.24×10^6 tonnes in 1982 to 1.96×10^6 tonnes in 1999 because of increased domestic demand.

With the limited potential for additional water supply, the pursuit of water sources for the improvement of water use efficiency has shifted to high-value crops. Vegetables and fruits are typically encouraged in China as economic agricultural products and as a way of raising rural incomes. The exports of vegetables and fruit increased from 1980-2000. This shift has produced high economic returns and made possible more selections of imports for the alleviation of the intensification of water stress.

Analysis of the Role of Virtual Water

Virtual Water Implications

Allan defined 'virtual water' as the water embodied in food crops that are traded internationally. Allan (1997) termed such food imports as 'virtual imports' due to the fact that they are equivalent to a transfer of water to the importing countries. The concept is used to describe a water-scarce nation's opportunities for achieving food security by purchasing a portion of its food requirements in the international market. It compares the amount of water embodied in a crop that can be purchased internationally with the amount of water which would be required to produce that crop domestically. Virtual water is a term that links water, food and trade (Allan 2003). As the metaphor of virtual water gained currency in the mid-1990s, economists at the World Bank started to use another term the 'water, food and trade nexus' (McCalla 1997).

This combines agronomic, economic and political outlooks, with emphasis on water as the key factor of production. The agronomic component involves the opportunity cost of water, which may include production of alternative crops or use in municipal, industrial or recreational activities. The opportunity cost of water use, which is a key component of the 'virtual water' perspective, must be considered when seeking an efficient allocation of scarce water resources. The same concept applies to the water embodied in other productive activities, such as manufacturing and power generation. Some manufactured goods require a greater amount of water in the production process than other goods. Nations in which water is particularly scarce may gain by importing water-intensive manufactured goods, while exporting goods that require less water in production. Similarly, nations with limited freshwater resources may gain by generating electricity in coastal areas, using ocean water for cooling, or by importing electricity from countries with abundant hydroelectric resources.

Policy Relevance of Virtual Water

Food trade plays an important role in redistributing the water resources of the world. The role of food imports in compensating water scarcity is elaborated with a quantitative accounting of the volume of virtual water embedded in their food imports. It is suggested that water be considered an economic good (Hoekstra et al. 2003a,b), and virtual water trade be seen as an instrument to achieve food security and efficient water use (Hofwegen et ai, 2003). However, when water policy issues are discussed, further implications of virtual water can be considered. Food imports are used when policy makers care about food security, while the feasibility of food imports is judged by opportunity cost. Wichelns (2003), however, thought that the optimal trading strategy could be determined by examining the comparative advantage, not the opportunity cost. This comparative advantage includes consideration of the water resource endowment and the production technologies in two or more countries that can engage in trade.

On the other side, virtual water has geopolitical implications. Trade induces dependence between countries, and it is influenced by and has implications on tariffs. Indeed, the issue of virtual water is related to globalization, which raises concerns among many politicians and the general public. Increasing global trade implies increased interdependence of nations (Hofwegen et al. 2003). Geopolitical conditions in food trade are not ignored because food security, like water security, is a highly contentious issue that needs carefully examining in terms of the degree and method of food self-sufficiency. Domestic political power has decided traded objectives, volumes and formations (Rang 2000). Bouwer (2000) suggested that an OFEC (Organization of Food Exporting Countries) be established with international controls, and in which the importing countries are ensured that global distribution of food will not be used as a political weapon. This study emphasizes the potential role of virtual water trade on the solution to water resource scarcity and food security.

Virtual Water Equivalents in Traded food

Food production is constrained by water scarcity; thus, importing food compensates for scarce water resources. Virtual land is also evaluated using imports divided by yields per hectare in the same year from the FAO statistics, where virtual water is the amount of imports multiplied by the global average of virtual water in every crop (Hoekstra and Hung 2003). On average, 8.9×10^6 tonnes of wheat and 0.9×10^6 tonnes of maize were imported annually between 1980 and 2000. An average of 10.20×10^9 m^3 water and 27.7×10^6 ha of land would be required to produce this imported wheat, while an average of 0.32×10^9 m^3 water and 1.6×10^6 ha land would be required to produce the imported maize. The sum of 10.52×10^9 m^3 of virtual water is equivalent to 23% of the annual average transfer of water volume for the South-North Water Transfer (SNWT) project. From 1995 to 1999, net virtual water import volume of crops in China was calculated at 101.9×10m, and China became the fifth in the top-30 table of net virtual water countries (Hoekstra and Hung 2003). The sum of 29.3×10^6 m^3 of virtual land is equivalent to 19% of crop arable land in 2000.

Water Compensation Through Virtual Water

Water resource scarcity constrains food production, and food supply cannot completely depend on the enlargement of food production. The basic food supply comes from food production and imports. Food import is appropriate when food price is lower than domestic production cost in the international market (Allan 1999). For virtual water embedded in the cereal imports, an aggregate ratio of 1 kg grain to 1 m of water is used. As such, the figures for cereal imports can be transformed into the same numeric volumes of virtual water in m/per capita. Research shows that virtual water volume embodied in cereal is higher than the global average (Oki et al. 2003). This means that imports of virtual water in China should be greater than estimates in this study. The factors that influence opportunity cost of production caused by the growth of irrigation water volume to enlarge the food supply have two aspects. One is the food supply capacity including food demand, food alternatives, food productive capacity and native import capacity. The native import capacity includes international food prices, native foreign trade conditions and international aid. The other aspect is concerned with water resources, embodied in water use efficiency in agriculture, agriculture management, water resource volume, water pollution, wastewater treatment and reuse. Those limiting factors are complex and, when combined, the opportunity cost in different nations and regions are obviously different. The amount of agricultural produce derived from a certain amount of water is subject to a large number of variables. These include the geographic location of the region where crops are grown, the irrigation system used, the management of the irrigation system, soil types and climatic conditions (Turton et al. 2002).

Moving virtual water will be much cheaper than moving water itself (Bouwer 2003). Global movement of food from water-rich to water-poor countries should be just as feasible as moving petroleum products from oil-rich to oil-poor countries. Virtual water opportunities for saving local water resources include imports of electric power from areas with more abundant water for cooling of thermal power plants, or with dams for hydroelectric power production (Bouwer 2000). Currently, most researchers in China think food demand and supply in China are on firm ground. Some agreed that the food import dependence ratio might be augmented to 20% (Wang et al. 2002), while others argued that food supply should be self-dependent to ensure food security and food trade between regions within China (Cheng 2003).

Virtual water is useful to alleviate water scarcity pressure, but many governments are worried that over-dependence on other nations will result in a food crisis and social instability. Importing virtual water embedded in traded food in China can be seen to alleviate water stress, and even achieve food security. Virtual water trade may compensate for water demands of not only the past but also the future. Meanwhile, the trade can store water in its virtual water form, especially by food storage, to play a potential role in solving food problems. However, policy considerations of domestic water and food security are needed to promote better understanding of the measures that need to be taken to ensure their future supply.

7

Complex Attachments to Water-places

This chapter explores water as central to defining complex attachments to place, an approach that is concerned to contribute to discussion beyond water's conceptualisation as a vital and natural resource in need of environmental management. The value of an active discourse and practice that emphasises water as a resource to be managed is obvious, for example, in cases where a sudden increase in water flows has the potential to transform rivers into floods with serious consequences for humans and other species. My interest, however, is to consider whether discussion of attachment to water sources in the Kimberley region of northern Western Australia, can show how, and to what extent, contemplation of water/human relationships might fruitfully facilitate understandings of 'place-based cultures' (Escobar 2001:142).

While notions of 'attachment' are sometimes conflated or confused with concepts of 'belonging'—about how and why persons (and/or native and introduced species, and so on), reveal or describe a sense of belonging to certain places and environments—the emphasis here is primarily on attachment. Like my co-contributors (some of whom examine concepts of belonging), I am concerned to extrapolate varied interpretations about a key environmental trope—in this case, water—to foster a deepened understanding not only about a precious environmental resource, but also about social life, cultural politics and material struggles in an Australian setting.

Concentrating on the remote Kimberley region's Fitzroy River, and examples where water sources sit on their own or become agents of transformation (for example, as rain or in food preparation), I begin with a consideration of guiding themes and then turn to Australian literature about people's relationships to water, especially to texts concerned with indigenous perspectives. A selection of Kimberley ethnographic data, a region where I have undertaken fieldwork since 1981, follows. Discussion of this material is used as a backdrop to canvass the ideas of Escobar (2001) on attachments to place, and of Tsing (2002, 2005, 2007) on the creation and outcome of 'friction' when cultural and environmental perceptions and activities collide. I conclude with the claim that studies of human/water relationships benefit when they are examined within the context of a place-based-culture analysis. Such a context is characterised by a mix of attachments that happen to be about places that involve water.

Friction, Attachment, Place

Conflicts and disputes over scarce or abundant water sources are major points of inquiry in current water studies. Tsing's use of 'friction', however, has the potential to provide a more comprehensive and nuanced way to explore the complexities of disjuncture and/or resonance that occurs when proposed developments threaten fragile ecologies and the life styles of nearby populations. This is especially the case on projects involving indigenous groups. Tsing (2002:472) encourages intellectual and ethical space to help make visible, rather than leave in unexamined form, problematic ideologies about environmental and indigenous interests. Tsing adopts the expression 'friction' to encompass elements of what she describes as the 'grip' and 'irritation' that interact to, on the one hand, facilitate and, on the other, limit, productive engagement and positive outcomes when environmental matters are a point of departure (Tsing 2005, 2007:58).

Tsing's (2005, 2007) work is mostly focused on Indonesia, but her conceptualisation of 'friction' facilitates a fertile framework from which to examine issues outside south-east Asian locations. Friction also sits nicely alongside 'traction'. Whilst arguably a term that emphasises the value of 'grip', traction is increasingly used in Kimberley settings to indicate a desire for concrete commitment and action. The comments of a local indigenous resource agency coordinator (identified here as 'Leo') give voice to my claim:

What people [referring to indigenous communities] around here need is a bit of traction—government and industry getting a grip on what people are talking about, what their needs are. Consultation means nothing around here without grip, without proper traction, without something or someone actually taking hold (Fieldnotes, Kimberley, November 2007).

Leo's use of traction is not too distant from Tsing's 'friction', albeit in contrasting contexts. His comments, made to myself and two indigenous people present, took place after a phone conversation with a major developer who had offered to put a large amount of funding into local initiatives—as long as any plans that were put forward accorded with the developer's time-line.

In later discussion, it became evident that both the phone call and the caveat were not unusual occurrences and, despite a need for funding, the offer had implications attached to it that were not straightforward, a point that indicates how Tsing's use of 'friction' might be understood.

This is most apparent when Tsing argues that evidence of 'grip' and 'irritation' as 'friction' emerges when examining social and environmental movements and their relationship to the corporate sector. In the above case, for instance, it is obvious that whilst the indigenous organization was offered some sort of 'grip' in the form of financial support, the offer was not without the 'irritation' of an unrealistically short timeframe. As Tsing notes, however, while there is little evidence to show that social, environmental or indigenous movements have the ability to 'displace the hegemony of private property and capitalist development', a certain 'messiness' (2002:472, 476) exists in such situations, such that 'not all indigenous peoples support environmental causes; some are in active conflict with conservation' (Tsing 2007:55). Escobar (2001) seems less concerned with the problem of inflexible categories and the characterization of issues that lend themselves to an analysis based on friction.

His concern is on practical engagements with, and resistances to, the unwanted development of places people hold dear for a variety of culturally endowed meanings. Escobar (2001: 147) is critical of studies that over-emphasise mobility and de-territiorialization to the extent that 'place-based practices and modes of consciousness for the production of culture' are overlooked. A proponent of the need to problematise unqualified generalisations, he also aims to de-centre the uncritical use of 'Western' forms of knowledge. Concentrating on the work of theorists such as Auge (1995), whose binary construction of 'non-place' tends to undermine concepts of place as multi-sited, contingent and relational, Basso (1996) and Bender (1993), among others, Escobar claims that contestations about place help to unravel the tensions embedded in material struggles. Such a process inevitably attracts consideration of local cultural beliefs and practices (2001:170171). Adding a further layer of meaning with regard to how concepts and constructs of place are constituted, Escobar cogently claims that local knowledge can also be read as political knowledge: 'Local knowledge is a place-specific way of giving meaning to the world'.

Whilst commending the literature on place as a site of sensory and phenomenological attachments, Escobar's main concern is with the politics of place-based cultures. He shows how attachments to place are constantly created and re-created by people's social and political engagement with that place, such as via active resistance to protect an area against development projects predicted to cause harm. In Escobar's parlance, social, cultural and political attachments are formed and become meaningful through active dialectical processes of 'connectivity, interactivity and positionality'. Such an approach logically includes consideration of divergent identities that emerge within 'the dynamics of place, networks and power'.

Giving prominence to the need for the cross-fertilization of political, economic and cultural approaches within the theories and practices of a political ecology framework, Escobar (2001:153) opines that an inter-mix of attachments to 'place' is given meaning by diverse groups of women and men through active political involvement. Escobar is also concerned to analyse attachments to place as an outcome of beliefs and practices where interactions between persons and their environments are in the constant process of mutual transformation. Such a process accords with Tsing's view that social (including indigenous or 'subaltern' to use one of Escobar's terms) and environmental movements fluctuate over time, a process that results in them being difficult to classify coherently, especially in circumstances where the local and global interconnect.

Read separately or in tandem, what these authors provide is a creative and dynamic framework to explore how and when description of water/human relationships might benefit from an approach that privileges culture as place-based attachment. That social, indigenous and environmental movements and issues can be constructively interpreted through the intertwined notions of friction and traction, is also a point of inquiry.

Next I outline a selection of Australian literature in which it is plain that contemplation of water prompts a rich and diverse range of studies. Most of these texts concentrate on Aboriginal and Islander groups and inter that water-places are significant religious and economic sites, sources of survival, and places endowed with cultural and political meanings.

Human/water Relationships

A visible social science and humanities research and publishing interest in water/human relationships has become increasingly evident in the past decade. Such a key epistemological shift has the potential to ensure that cultural description and analysis, and past and present human activities, is regularly embedded in studies of the environment. It remains the case that more independent anthropological and cross-disciplinary research needs to occur to augment environmental, biophysical and natural science inquiry, but there is no doubt that a substantial volume of more explicit culturally attuned work with water at its centre is emerging. As well as contributions in this volume, monographs in the field of Australian, Pacific and south-east Asian water studies include those by Balint (2005), Brearley (2005), Leybourne and Gaynor (2006), and LahiriDutt (2006). This list does not include works outside these geographic areas, nor does it acknowledge texts focused on relational inquiries, such as that by Beresford et al. (2001) who discuss the salinity crisis in Western Australia from social science perspectives. Recent trends are especially evident in research being conducted with and/or about Aboriginal and Islander Australians, in part because of the High Court's 1992 Mabo Decision where the Meriam Islanders in the Torres Strait off the northern Queensland coast were found to have long-term cultural affiliations with not only land but also with fresh and saltwater sources on the Island of Mer. Altman (2004), with regard to water issues, canvasses property rights and interests in native title provisions and land-use agreements, and Bagshaw outlines the Kimberley Karajarri native title claim where water featured prominently.

Notwithstanding the water-based research that is now occurring in native title claims and cross-disciplinary environmental research more broadly, a selection of material involving Australian indigenous groups reveals a spectrum of relational topics ranging from past and present subsistence and customary law marine activity, to spiritual affiliations, beliefs and rituals, to environmental implications of water planning policy. Of particular interest is that each writer indicates, without always making it explicit, water-inspired beliefs and practices as these accord with the making and meaning of place-based cultures. Bayly (1999), for instance, brings anthropological, archaeological and historical data together to show the ability of indigenous groups in the Western Desert bloc to 'read' the landscape and find surface and groundwater in desert locations, a quality that disadvantaged non-indigenous travellers and explorers in the distant and recent past whose cultural experience did not involve such engagements.

Like other commentators and researchers, Bayly argues that, historically, indigenous migration was largely dictated by knowledge about the presence or absence of water, and water's ability to transform the environment. Rose (2004) is also concerned with how water sources were and are central to sustaining life and land, especially in the desert during prolonged drought. Rose, in a way comparable to Escobar's (2001) usage, describes this process as one of 'connectivity' that includes ground water, surface water and rain mediated through the Rainbow Serpent or mythical snake deemed responsible for water's creation. As Rose makes plain, indigenous groups had in the past and continue in the present to have a responsibility to care for all water sources. Any negative interventions (such as the uncontrolled drinking of water by cattle, or the implications of mining, both of which can result in contamination) pose major problems for people (indigenous and non-indigenous or 'settler') and the culture-based landscape in which they are embedded.

Writing specifically about the Gulf of Carpentaria in northern Queensland, Trigger's (1985) research takes a more fine-grained ethnographic approach to explore the naming of fresh and saltwater sources. He argues that water is a highly significant feature of the environment: water acts as a 'key signifier of meanings attributed to nature that extend beyond water places themselves to encompass the landscape more broadly'. Langton (2006), with supporting cultural data gathered from groups with traditional affiliations to the Laura Basin area of Queensland's Cape York Peninsula, claims that people relate to water through the regular and interconnected cycle of 'earth, wind, fire and water'.

Morphy and Morphy (2006), sharing some resonance with Langton's discussion of the relationship between spirituality and the environment, describe how successive generations of Yolngu at Blue Mud Bay in the Northern Territory's Arnhem Land enact rituals to enhance contrasting spiritual connections with salt and freshwater places. In a framework conceptualised as 'eco-mythology', where the authors are keen to make a distinction between interactions with land and with water, Morphy and Morphy discern that an imposed dualism between salt and freshwater 'fails to capture the complexity of the environmental, ecological and mythological relationship between the two' (2006:74). Morphy and Morphy also observe that 'The Yolngu have an overall model of water as process', a claim that is explored through the use of metaphor to explain fluctuating patterns of water ownership, ritual performance, local interactions and attachments to water-based places. Magowan (2007) writes too about Yolngu attachments to water in a way that highlights the metaphorical re-cycling of practical and symbolic interconnections. These are evident in description of water currents, long-term ecological knowledge, music production and performance, and an all-embracing spirituality derived from the body of religion and law known as the Dreaming (2007:133-141).

Whilst I am less concerned in this chapter with people's relationship to saltwater sources and marine tenure than with freshwater rivers and related waterways, Bradley's (2006) research is useful here because it attends to how (salt)water influences identity formations and kinship interactions among the Yanyuwa living in Queensland coastal communities. In such a process, attachments to (salt)water-places are privileged. His work cogently embraces the sociality and interdependence of humans with their environment, a summation that specifies cultural attachments to saltwater places that have been reproduced throughout time.

According to Bradley, identity is constructed around named and relational saltwater sources and interdependent sea-living species. Similarly to other authors whose work, Bradley presents rich ethnographic data to consider water/human issues as these explain Yanyuwa social and cultural life. He argues that the Yanyuwa see no distinction between land and water and that 'the sea and seagrass beds... [are all part of]... geographic land units' (Bradley 2006:128). Indicating the importance of saltwater interactions, alongside analyses based on the qualities of 'sea knowledge' and attendant power relations, Bradley concludes that Yanyuwa identity is largely founded on 'enduring emotional links between people, sea-country and many of the creatures which inhabit the area'.

Goodall (2002) contemplates issues of water and identity too, and in a way that accords with Escobar's (2001) place-based cultures framework. Her research also facilitates a return to freshwater rivers and associated waterways, in particular how forms of attachment were heightened when water resources were threatened. Focusing on indigenous and non-indigenous (or 'settler') groups in New South Wales who contested irrigation for cotton production on the Darling floodplain, Goodall shows how the project impacted negatively on local groups and the broader social and political life of nearby towns. Cognisant of the intertwined relationship between land and water, she observes that people construct attachments to place because of their (often long-term) lived experience with it, their social history and livelihood, and a series of remembered, sentimental associations. In such places, water is a key social as well as environmental element because 'it is water which creates and replenishes the land and the soil' (2002:36), a circumstance that influences people's work practices and social activities at nearby lands and waters.

A further dimension to Goodall's research is that demographic changes are shown to have occurred as a result of introduced irrigation and cotton production. Demographic change has obvious implications for social change. In the case presented by Goodall, one of these was that many senior citizens moved away from the area once the development was established. Their relocation resulted in the loss of long-term local knowledge, support and experience (2002:47-48). Goodall also shows how people's hitherto positive attachments to places associated with the floodplain were frequently replaced by 'a set of fears so deep they can only be expressed in terms of overturning the most basic rule of nature'. The struggle over a place around which water was central to how people constructed and maintained their identity, enacted social relations, undertook past and present work-practices and sustained sentimental associations, resonates with Escobar's ideas on how and when attachments to place are created and re-created. Evident also in Goodall's material is that indigenous groups and non-indigenous graziers were united in their opposition to the project, and that friction existed between them and the developers. This situation also reveals that attachments to place were amplified when the Darling floodplain was threatened; whilst the floodplain mattered, it was the struggle over it that mattered most.

Among the issues that emerge when evaluating water/human relationships in Australia, is that most studies highlight the intrinsic, local nature of the land/water/people interaction and connection. It also becomes clear that salt and freshwater sources should not be treated as one and the same. That a hydro-and ecological change in how a water source is used can lead to a change in how people relate to it encompasses a range of culturally complex issues, including that water is engendered with a variety of meanings. Identity formation and kinship affiliation can also be determined through research on water, as can knowledge about contested usage, and patterns of migration to and from temporary and permanent water places. Each study implies, and sometimes explicates, ways in which local groups become attached to sources of water beyond water's obvious nourishing, life-giving force. This is especially the case when water sources are endangered, and cultural ideas, beliefs and activities collide. The Kimberley's Fitzroy River and associated waterways present excellent scope to evaluate matters related to how and when attachments to place are formed in the way Escobar describes. It also presents data to explore Tsing's conceptualisation of 'friction'. I begin by explaining a little of the Kimberley, in particular the Fitzroy Valley in the West Kimberley landscape, and then turn to a case-study which features the Fitzroy River.

Kimberley Water-places

The Kimberley is a vast and rugged region of over 100,000 square kilometres with temperatures ranging between 20 and 45 degrees Celsius and landscapes alternating between desert, ranges and coastline (Brown 2004; Storey et al. 2001). A place of many contrasts, the Fitzroy River and its tributaries (including the intersecting Margaret and Leopold Rivers) is central to understanding the West Kimberley landscape from different environmental, geographic, economic and cultural vantage points. One of the last unregulated rivers in Australia, the Fitzroy River traverses from the West to the East Kimberley across 85,000 square kilometres of landscape (Storey et al. 2001).

Published and unpublished research in the field of native title discusses indigenous affiliations with a range of Kimberley water sources, such as rivers, soaks, creeks, pools, lagoons and 'living water' (known locally as jila) and so on. Research available in other material describes the manifold ways in which groups manage and make meaningful interconnections with waters and rivers. In the West Kimberley, the Ngarinyn, Karajarri, Bunuba and Gooniyandi are renowned as river owning and using groups, with members of the desert-related Walmajarri and Wangkajunga groups establishing a form of cultural co-existence in the area as the result of a 40 year process of voluntary, involuntary, and negotiated migration. A Walmajarri woman, Daisy Andrews, describes this interaction thus: 'It's really good now how we can all sit down and work together. It's really good, river and desert together, whole lot' (Mangkaja Arts, 2003:6). Other desert people evoke river-based totemic affiliations for children born post-migration, and often describe how the river people taught them how to fish. Walmajarri woman 'Ngunda' put it this way:

We had to learn about the martuwarra [river], how to fish with string and burnt needles shaped like a little hook, we didn't have throw nets. Nothing! The Gooniyandi and Bunuba [meaning river groups] taught us how to fish—how to hunt, cook and eat fish (Fieldnotes, Kimberley, July 2006). In parallel with other Australian indigenous groups and elsewhere, distinctions exist between salt and freshwater sources, and rivers and other forms of water (such as rain, and permanent or ephemeral waterways), continue to be conceptualised as having been inspired by the Dreaming. Home to approximately 40 Aboriginal communities and four major towns, people in the region, like the landscape of which they are an integral part, are constantly responding to change and some extremely tough social and economic conditions. Despite these conditions and, in many cases what can only be described as circumstances of poverty and neglect, indigenous rights to, and interests in, land and water continue to be based on a complex combination of patrilineal and matrilineal descent, conception sites, birth sites, totemic affiliations, historical associations and residence. Whilst economic initiatives for indigenous groups are improving in the region, evidenced in projects such as the Manbana Aboriginal Aqua Culture Centre in Broome which is now providing study programs for local people, aquatic tours for visitors, and expanding fish markets into south-east Asia, many Kimberley groups continue to exist on government controlled social security entitlements, an economic problem with social implications that has received ongoing local attention. Opportunities that add to household incomes include contract cattle station work, intermittent involvement in tourism activities, for example as tour guides on Aboriginal land and in national parks, and the results of subsistence work.

Fishing is an important economic activity too, including the exchange of fish when one family is more successful than another after a day at the river. Education for young people also occurs via fishing with the teaching of local language names for fish during what would otherwise be regarded as recreational and economic activity. During periods of mourning when fish as a sole food source are necessary to accommodate taboos on the consumption of beef products following the death of a loved one, the catching, cooking and consumption of fish becomes vital (Toussaint et al. 2005:66). Relationships to water also continue to be understood by most women and men as having been inspired by the Dreaming, known locally as Ngarrangkarni or Jumungkarni. Walmajarri man, Joe Brown, recounts the cyclical interconnections between the Dreaming, people, land and water:

> *Everything we know comes from the Dreamtime. People even in town still know how to sing story for their country. Every jila [water hole] has its own songs, story and skin groups. Without the snake underneath the water will go away. We have been*

> *looking after our waterholes and rivers for thousands of years. We have respect because we know that if you don't treat it right many things can happen. This is the reason that we need to make other people learn (quoted in La Fontaine 2006:217).*

Janjin Rogers, also a Walmajarri man, adds example to Joe Brown's description. His narrative also reveals the transformative and cyclical qualities of water:

> *My mother was worrying about me getting thirsty. She started singing for rain, I couldn't believe that the rain would come. There was a little cloud in the sky, my mother sang it and it got really big. Then it started pouring rain on the ground, water everywhere. I was really happy. My father came back, he saw everything wet and he was happy for the good rain (Karrayili Education and Manghkaja Arts 1993:8).*

Seasonal rains, whilst sometimes causing flood and community isolation for up to three weeks a year, are central to ecological restoration and to local perceptions of what is considered to be the river's natural course: annual monsoonal rains ensure the replenishment of vital resources (waters, lands, plants, animals, fish, and so on) for humans and other species. River stories also find expression in the production of art, such as paintings made for sale to locals and visitors, and for state and national art exhibitions (Toussaint et al. 2005:68-71). Paintings about water also increasingly reflect political themes, including the need to protect local rivers and waters with which people have long-term culture-based connections and attachments. This point is elaborated below. From the insights raised here and expanded elsewhere, it can be inferred that Kimberley indigenous stories and activities regularly explicate complex matters relating to how people interact with, and subsequently form attachments to, the river and other water-places. The water/human relationship evident for Kimberley groups also shares certain parallels with indigenous peoples in other northern Australian settings. Local groups have rights and interests in waters and lands through customary law, via forms of residence, and as a result of desert migration. The painting of water-places has added to some people's income, and the river presents a location where subsistence activity and learning occur. Less well known is how non-indigenous people relate to local water-places. A discussion of non-indigenous/water relationships is, of course, just as vital as those of an indigenous kind, especially when canvassing how frictions might emerge and place-based cultures be identified.

The Fitzroy River, Friction, and Place-based Attachments

Unlike indigenous peoples who have generations of involvement with the river's creation and protection through their own body of religion and customary law, social and political networks and economic activities, settler relationships are less well documented. In broad terms, these can be classified as having occurred more recently and incrementally through temporary and permanent residency, often involving work for government, non-government and industry (e.g. tourism, mining), and pastoral work on cattle stations. In addition to work-related interactions, activities such as fishing, camping, bird watching and recreation are regularly reported.

Such practical, economic and social activities have undoubtedly established memorable ties of affection and sociality that, on some occasions, are shared with indigenous people. Settler environmental groups are the focus group here, especially as they have been actively involved in protecting the river in particular, and the broader environment more generally, against unwanted developments. This circumstance has drawn a mix of groups and organizations together. A timely beginning is 1996 when government and industry plans to dam the Kimberley's Dimond Gorge, and two of its major tributaries on the Margaret and Leopold Rivers, became public knowledge. With parallels to the study outlined by Goodall (2002), the aim was to establish a cotton enterprise based on irrigated water. Local organizations such as the Kimberley Land Council and the Kimberley Aboriginal Law and Culture Centre contested the development with support from national environmental agencies such as the Australian Conservation Foundation. As I have shown elsewhere (Toussaint 2008a), this situation united local groups in their opposition to the proposed developments. The regional aim, involving a diverse range of local indigenous, tourism, hospitality and pastoral groups, became to protect the Fitzroy River's waters for Kimberley people and Kimberley lands, in particular the economic value of locally-controlled tourism to the region, of which the river was a central component.

The 1996 initiative also resulted in the establishment of a Kimberley-based environmental organization that involved non-indigenous people committed to local conservation and nature issues. Called 'Environs Kimberley', the organization has become increasingly active in the region where it now has a Broome-based office, several permanent staff, and a large assortment of dedicated members and volunteers.

Whilst the 1996 plan was eventually withdrawn, in 2005 the Western Australian State Government (in concert with industry) again expressed interest in the Kimberley's extensive river system. They established an inquiry to assess the feasibility of diverting the river and/or transporting its water 3,500 kilometres south to offset water shortages in the State's southern region. Three options were investigated: a Kimberley to Perth pipeline, a canal and ocean transport. The inquiry recommended in its final report that all three options were too costly and that limited knowledge existed about the 'hydrogeology and ecology' of the West Kimberley. Despite such an outcome, indigenous and non-indigenous groups continue to be agitated that some form of river diversion is not too distant from the minds and plans of government and industry. This has resulted in a series of nuanced frictions. Before explaining this situation, it is useful to outline certain beliefs and practices that characterize non-indigenous relationships to the river, some of which parallel those of an indigenous kind.

'Sam' has lived in the Fitzroy Valley for several decades. He is one of the few non-indigenous workers regularly employed by an indigenous organization. His work often involves him in field trips to desert locations several hundred kilometres south of the town of Fitzroy Crossing. Sam's comments make plain his love of the area, especially the Fitzroy River:

> *I like going to the desert [hundreds of kilometres south of the Fitzroy River] for native title and heritage work. We usually travel in convoy, a few four-wheel drive vehicles travelling together. It's the jila [water holes] I'm most interested in. And then I'm happy to be back [in town] near the river. It's everything here [Fitzroy Valley]. I wouldn't be here if the river wasn't here. A lot of people wouldn't be (Fieldnotes, Kimberley, June 2006).*

Sam's views can be positively compared with those of a local ecologist who worked for an indigenous resource agency and described himself as feeling 'passionate' about the Fitzroy River, a place where he regularly fished and camped on weekends. Noting the importance of the river to his work, 'Didier' commented that frictions could arise 'when indigenous people aren't actually present to explain their interests to outsiders, such as to developers...' (Fieldnotes, Kimberley, July 2006). Despite indicating a passion for the river, Didier also stressed his long-term 'commitment' to the place and the local people, in part by talking about how 'proper, fee-paying' tourism might help to support and conserve the Kimberley's ecological and cultural environment. He commented on the importance of exploring the feasibility of commercial projects that might:

> *result in the environment being more sustainable, and help to get people and organizations away from a reliance on government funding. The traditional owners [meaning local indigenous people who sustained customary law rights and interests in lands and waters] would need to agree, of course, and their rights and interests in all this [lands, waters, coastlines] are not unreasonable. But we've got a long way to go, and not many resources.... (Fieldnotes, Kimberley, July 2006).*

In such a positioning, Didier makes plain his concern for indigenous groups and cultural matters. But in a later conversation, he positioned himself slightly differently. Didier pointed out that whilst it was constructive to see that indigenous people were becoming increasingly involved in carefully controlled tourism projects (such as at a place called Jarlmadangah where several Walmajarri and Mangala families operated a successful tourism business), he and other ecologists were worried about the impact an increase in river cruises was having. Run by another indigenous group with assistance from local operators and/or in conjunction with state-based conservation agencies, the boat cruises for tourists had resulted in the Fitzroy River experiencing expanding environmental flows, and the wash from these often impinged on the river's banks with consequences for wildlife. For example the nesting habits of crocodiles were affected as they had been obliged to move further inland (to less safe places) when the seasonal time came to lay their eggs (Fieldnotes, Kimberley, July 2006). Of interest in this scenario is that while Didier's comments suggested he could understand the need for indigenous groups to be involved in a tourism venture, he felt torn about such an initiative because of the venture's impact on the environment. What was not at issue, however, was his attachment to, and connections with, the Fitzroy River as a culturally-endowed place. Whilst it is evident that non-indigenous people such as Sam and Didier value the Fitzroy River and the broader environment, this was not always the case. 'Craig', an environmental health officer with a government agency, had quite a different perspective:

> *I don't really care much about the environment here, like it's interesting, I guess, and I'm an environmental health officer so I get out and about a bit, into the bush, along the river system, visiting communities and stations... but it's the easy life style that gets me in. It's nothing to do with the environment, and I'm not sure how long I'll stay here (Fieldnotes, Kimberley, June 2006).*

Craig's comments make clear his position. At the same time they both support and diverge from Escobar's framework about how and why attachments to place occur. At one level, Craig showed that he is attached to the environment because of the 'life-style' it provided, but at another he revealed the lack of a long-term commitment to people and place in a way that differs from Sam's and Didier's emphases. The comments of a freshwater campaigner, 'Gary', are particularly apposite here, perhaps because they help to crystallise the two perspectives. With a background in sociology and ecology, Gary's work with a Kimberley environmental agency shows how his attachments to two different rivers—one of which is the Kimberley's Fitzroy River—were reliant on a struggle to protect each resource. His personal positioning clearly shows how human/water connections were established through active political engagement:

> *I can see that it [the Fitzroy] is a beautiful river... but I haven't formed an attachment in the way that I did for the McArthur*

> *River [in the Northern Territory]. I was a freshwater campaign officer there too and although I only visited the river twice, I formed a real attachment to it... I felt passionately about it, partly through my experience with the local people... but also because I had to defend the river against mining and other activities that threatened to damage it. I guess you could say I had a personal experience with the McArthur... I might develop something with the Fitzroy River if I have to campaign for it. It depends a lot on what happens when I do more work on it and mix with the local people.*

Coordinator of Environs Kimberley, Maria Mann, makes a similar point about how extensive involvement with a place can enhance people's social and cultural relationship with it. In a 2006 event celebrating ten years' of active campaigning, Maria was on site near a narrow Fitzroy River Gorge when asked by a journalist to talk about river campaigns: 'You have to try to imagine—if the dam had gone ahead this would all have been under water...' (quoted in Prior 2006:4). In the same article, co-founder of Environs Kimberley, Pat Lowe, said that the environmental group:

> *simply couldn't believe that it [damming or diversion of the Fitzroy River] would happen in the Kimberley... There's been so much land clearing and so much destruction already... Also, we were concerned about riding roughshod over indigenous peoples' rights (quoted in Prior 2006:5).*

After a three day event at a Kimberley Wilderness Conservancy Camp, about 50 Environs Kimberley members were involved in activities such as 'bird watching, ecology tours, canoeing, and swimming' (Prior 2006:5). As a parting comment, Lowe pointedly claimed: 'When we were flying out here Ito the Wilderness Camp]... I felt moved to look down on the river and think we had a bit to do with saving it...' (quoted in Prior 2006:5). What these comments reveal from non-indigenous perspectives, at least, is that when a water-inspired place became a contested site, people's struggle with it, their positioning with the river as ally, resulted in a profound sense of attachment. Such an attachment cannot, of course, be confined to settler groups. As I have argued elsewhere and above, indigenous groups have for countless generations sustained a complex suite of connections to, and attachments with, the Fitzroy River and associated water-places. A perfect example to illuminate this point further can be seen in the imagery and artistry of Gooniyandi artist Butcher Cheryl who, each time the Fitzroy River was threatened, produced a series of paintings to visually position his concern. A quote from Butcher articulates his concerns as well as his cultural connections:

> *I have been making these things (images) for a long time. I put what I am looking at, tree, river, bush tucker, all that thing, that Ngarrangkarni (Dreaming) story too.... I know that Ngarrangkarni from nay mother's side. I know that country too from long time, like Manjoowa [permanent waterhole on the Margaret River]. There's crocodile, fish, everything there. I know this country from long time, they can't change that river country [referring to the damming or diversion of the Fitzroy River], it's no good that idea.*

Whilst both indigenous and environmental groups remain concerned that efforts to dam or divert the Fitzroy River's flow have not completely disappeared, these are not the only issues they have to face. Others include the impact of large-scale gas projects, poorly managed tourism, the problem of introduced weeds, and climate change. Within this spectrum of concerns a number of frictions emerge. For example, INPEX, an internationally owned petroleum company is hoping to secure access to locations such as the Maret Islands off the Kimberley coast to progress a major gas project, a move that has distressed conservation groups. INPEX executives have also made plain their intention to directly consult with indigenous groups in keeping with an Indigenous Land Use Agreement being negotiated in accordance with policies embedded in native title processes. Such a possibility has upset local environmentalist groups whose interest is to protect local biodiversity and the broader landscape; it also signals the kind of friction embedded in Tsing's analysis. I can see that it (the Fitzroy) is a beautiful river, but....

With different forms of water as guide, and the culturally influenced human/environment relationship as imperative, it is possible to engender ethnographic description and analysis as a means of interpreting place-based cultures. From the data provided here it is plain that Kimberley indigenous river and desert-related groups have long-term religious, socio-cultural ties to water-places, and that these are central to understanding the broader historical ecology as well as native title laws. It is also the case that the Fitzroy River, in particular, is a meaningful place for many non-indigenous women and men. The river is central to why many people live, work, stay and play in the area.

Escobar (2001) claims that places are endowed with meaning because of what occurs there, especially in situations where cultural politics are at issue. In the case discussed here, attachments indigenous groups have with the river, alongside attachments that non-indigenous groups have developed with the river over time, result in the Fitzroy River being not only an important water-place, but also a site of meaningful engagement, including struggle over its usage. Sectors of indigenous and settler societies who live in the Kimberley care deeply about the river for many different socio-cultural, economic, emotional and political reasons: the river is part of their permanent or temporary homeland, some people have customary law affiliations with the river, others fish and swim in the river for food and/or for fun, and some cherish the times when the annual flood replenishes wildlife.

The sort of experiential knowledge and practice that influences people's history, beliefs, activities and passions about water-places that contain, channel and sometimes transform fresh and saltwater sources, provides a unique trajectory to understand what socially, politically and emotionally connects people to place. This situation emerges when water sources are potentially or actually affected by unwanted change or developments that threaten, and generate friction, about its place-based cultural value. That sources of friction, and forms of attachment, can be understood as complementary presents a framework that might help to extend knowledge not only about the water/ human relationship, but also about how place-based cultures are given meaning and by whom. This scenario cannot be solely confined to Kimberley indigenous groups. For some of the reasons I have described, non-indigenous peoples have also developed complex attachments to water-places overtime. One of the implications from this discussion is to inquire whether it mattered that the sites of focus were specifically marked as water-places. I have suggested here that while the waters and rivers mattered, engagements about their protection mattered more. It was through multi-layered social and political engagements over environmental and ecological protection that place-based cultural attachments were formed.

Postscript

Contestations over Kimberley water-places and environments have been increasingly amplified in the electronic, print and cyber media. With saltwater places and nearby islands under imminent threat of gas project development, local concerns have intensified. While indigenous groups have finally secured 'grip' via native title success, many are now facing the 'irritation' that such success has engendered. Notwithstanding ecological concerns, native title leverage ensures economic support (such as work and housing) if the project is agreed to, but not if it is declined. With or without native title success, indigenous groups cannot veto such a project. Whilst in no way categorically oppositional, alongside indigenous concerns are those of a non-indigenous kind, especially for people aiming to privilege the rich biodiversity of local waters, wildlife, lands and coasts. Lines are being increasingly drawn and intersected in human/environment articulations that cut across social groups (indigenous, settler, migrant), disciplines (conservationists, ecologists, naturalists), industry (pastoralists, developers, tourist operators, agriculturalists), and government and non-government agencies. Not too far from each interest and claim, are social and political attachments that reveal culturally complex anchors to place.

8

Clean Water Act Programmes

In the last decade the National Marine Fisheries Service (NMFS) has issued an unprecedented number of endangered and threatened salmonid listings under the Endangered Species Act (ESA). These listings affect virtually every major watershed in the Pacific Northwest and are reminders that these water bodies and the species they support are in peril. Such results should come as no surprise, given that over 870 water bodies in the State of Oregon alone fail to meet water quality standards under the Clean Water Act (CWA). While degraded water quality is one of many factors leading to the decline and extinction of salmonid populations, ESA listings expose the failure of the Clean Water Act to achieve its congressionally mandated goal of fishable waters.

ESA listings also raise important liability issues for state agencies administering Clean Water Act programs like the National Pollutant Discharge Elimination System (NPDES) permit program, as well as private actors who must comply with the CWA and the ESA. State agencies responsible for CWA programs face an unusual predicament: they must comply with minimum federal CWA requirements, but their actions are not federal activities triggering ESA consultations. This uneasy duality arises because the Clean Water Act authorizes states to assume responsibility under state laws for water pollution control programs in lieu of programs admirestered by the federal government. One potential result of the CWA's peculiar cooperative federalism structure is that state authorities may issue to private permittees NPDES permits that comply with the CWA but not with the ESA.

This anomaly leaves private actors uneasy and uncertain about the duties and liabilities the ESA places on them and their state-issued permits. As more aquatic species are listed under the ESA, it is crucial for federal and state agencies to understand how the ESA and the CWA can complement one another to protect aquatic species and their habitat. In an effort to provide stronger protection for listed species and to create "greater regulatory predictability for States, Tribes, and the public," the Environmental Protection Agency (EPA), the Fish and Wildlife Service (FWS), and NMFS developed coordination procedures in several draft Memoranda of Agreements (MOAs) for approving and reviewing state permit and water quality standards programs.

Most notably, the draft MOA procedures rely on water quality standards as the vehicle to protect endangered species. To ensure that water quality standards are protective of listed species, the latest draft MOA calls for ESA consultations when EPA triennially reviews state water quality standards.

Without a similar federal action triggering mechanism in state permit programs, the draft MOA instead proposes to establish a federal fish and wildlife review of state-issued permits, which must also comply with water quality standards protective of endangered species. While EPA's efforts to take affirmative steps to integrate the complementary goals of the CWA and the ESA are commendable and potentially signal important change, the true measure of success will turn on the actual protection provided by water quality standards to endangered and threatened species.

This Comment evaluates whether EPA's proposed approach to species protection under state-administered Clean Water Act programs is an effective one. As the agency institutes these new procedures in state CWA programs, will ESA consultations ensure that water quality standards—the baseline for state-issued permits—afford endangered and threatened species the level of protection mandated under the ESA? Moreover, will state and federal agencies embrace these material changes to improve water quality in order to meet the needs of endangered species?

Part II of this Comment provides the statutory background of the CWA and ESA, explaining how these Acts can potentially promote species protection via water quality standards. Part III analyzes how EPA's proposed 1992, 1997, and 1999 interagency agreements with the Services aim to protect aquatic species in state NPDES permit and water quality standard programs under CWA authority. Then it evaluates the most recent MOA, challenging the capability of the national consultation and accompanying biological opinion to actually ensure that state water quality standards and state issued permits will not jeopardize listed species or adversely modify or destroy designated critical habitat. That Part then considers the agency's application of consultation procedures to its most recently delegated state permit program in Texas and questions the ability of current water quality standards to adequately protect species.

Finally, a look at the recent section 7 consultation on Oregon's water quality standards raises skepticism about the agencies' commitment to setting and enforcing water quality standards that protect listed species. Part V concludes that EPA's interagency agreement is an important effort to protect species under state CWA programs, because it materially changes water quality standards to function as the centerpiece of the CWA. In this new role, water quality standards could potentially drive the NPDES permit program and address the nation's grave nonpoint source pollution problem. Recent attempts to implement this vision in Texas and Oregon, however, illustrate administrative reluctance and scientific challenges in relying on water quality standards to meet all the needs of endangered aquatic species.

The Endangered Species Act and the Clean Water Act can Protect Aquatic Species

The language of both the Endangered Species Act and the Clean Water Act refer to species protection and viability as overarching goals. The ESA aims to protect and recover species facing extinction pressures and the ecosystems on which they depend. To encourage nationwide conservation, the ESA requires federal agencies to cooperate with state and local agencies "to resolve water resource issues in concert with conservation of endangered species." Similarly, the CWA strives to "restore and maintain the chemical, physical, and biological integrity of the Nation's waters," a more expansive goal than mere pollution control.

One measure of biological integrity is the degree to which the water affords "protection and propagation of fish, shellfish, and wildlife," another primary goal of the CWA. Currently though, discharges of pollutants authorized by federal or state CWA permits may adversely affect proposed or listed endangered aquatic species and their habitat, thereby undermining the objectives of both the CWA and the ESA. Despite the overlapping goals of these environmental statutes—achieving healthy aquatic ecosystems—the statutes have largely functioned independently of each other in the past. This Part examines how the CWA and the ESA independently address species protection and how ESA consultations regarding state water quality standards may serve to link these interrelated statutes and further the goal of sustainable aquatic ecosystems.

The Clean Water Act

Achieving fishable waters was a primary goal of the 1972 Federal Water Pollution Control Act Amendments, now commonly referred to as the Clean Water Act. To this end, Congress set out to eliminate the discharge of pollutants into navigable waters by 1985, to attain and maintain fishable and swimmable waters by 1983, and to immediately prohibit the discharge of toxic pollutants in any amount. As part of this grand water pollution control scheme, the CWA gives "due regard... to the improvements which are necessary to conserve such waters for the protection and propagation of fish and aquatic life and wildlife." Not only did Congress explicitly refer to fish and wildlife protection in the CWA's preamble, but also throughout its substantive sections. For example, the CWA establishes qualitative water standards that consider the use and value for "propagation of fish and wildlife" and quantitative effluent discharge limits that in turn must meet these water quality standards.

The Cooperative Federalism Structure of the Clean Water Act : The Clean Water Act created the National Pollutant Discharge Elimination System under section 402 as the primary mechanism to achieve its fishable waters goal and established water quality standards (WQSs) under section 303 to supplement this new water pollution control regime. The NPDES permit program regulates the discharge of pollutants from point sources into waters of the United States based on predetermined technology-based controls. By contrast, state water quality standards manage both point and nonpoint source discharges based on how they qualitatively affect the receiving waters. Under the CWA's cooperative federalism scheme, EPA and the states share concurrent authority to develop and implement water pollution controls. While the CWA does not require states to operate its programs, the statute's language and legislative history indicate a clear congressional intent for states to assume primary responsibility in "prevention, reduction, and elimination of pollution," while EPA was intended to play only a supervisory role.

State Assumption of the NPDES Permit Program : Section 402 of the CWA authorizes EPA to approve and oversee state-administered NPDES permit programs for point source discharges into waters. States seeking to administer an NPDES program must demonstrate that they have the legal authority under state law to operate and enforce the program consistently with the CWA's minimum requirements. The statute requires EPA to approve the program if the state meets all nine enumerated requirements listed in section 402(b) of the CWA.

The statute requires the state to provide public notice of all permit applications and submit copies to EPA. States also must submit draft permits to FWS, NMFS, and any other federal agencies with jurisdiction over fish, shellfish, and wildlife resources and receive comments from the Services on state-issued permits. Although EPA cannot refuse state assumption of species protection under the ESA or other federal statutes, the agency retains discretion under section 402(b)(1)(A) to incorporate effluent limitations and "any more stringent limitation[s]" under section 301 by reference. This provision means that when technology-based or secondary treatment limits are not adequate to satisfy water quality standards in receiving waters, NPDES permits must provide more stringent limits. Such limitations could include current state water quality standards protective of listed species. Once EPA approves a state's permitting authority, the agency suspends the issuance of federal permits. To date, EPA has delegated NPDES permitting authority to forty-three states and one U.S. territory and has never revoked state authority, even though it has the power to do so.

Approved states must ensure that newly issued permits comply with federal technology-based controls. These end-of-pipe effluent limitations restrict the type and amount of pollutants that may be discharged into navigable waters based on the source's technological capabilities. While federal effluent limitations serve as a minimum, states can adopt more stringent pollution-control standards. As long as permittees abide by applicable effluent limitations, states cannot impose new effluent limitations during the permit's five-year lifespan. Compliance with an NPDES permit means compliance with the obligations of the CWA, thus shielding dischargers from CWA liability.

State Water Quality Standards : In the CWA, Congress recognized that technology-based controls might fail to meet state water quality standards despite individual compliance with effluent limitations. Consequently, Congress required NPDES permits to include "any more stringent limitation[s]" necessary to meet water quality standards developed by the states. This requirement is aimed to prevent water quality from falling below protective levels for species in states that adopt such standards.

Water quality standards provide the strongest authority for species protection in state NPDES programs because these standards often set numeric goals for specific water bodies and are the foundation for water-quality-based effluent limitations for NPDES permits under section 301(b)(1)(C). Water quality standards seek to improve the overall quality of water bodies by 1) designating uses of water bodies, 2) setting criteria to protect those uses, and 3) creating an antidegradation policy that protects existing uses and ensures high water quality.

Water bodies unable to meet water quality standards qualify as "water quality limited segments," which states rank based on their importance and level of pollution. States then establish total maximum daily loads (TMDLs) for each water quality limited segment by identifying the maximum amount of pollutants a water body can receive from all point and nonpoint sources each day without violating water quality numeric and narrative standards. These levels, in turn, serve as the baseline for allocating pollution control responsibilities among point and nonpoint source dischargers.

States promulgate water quality standards, subject to federal approval and review on a triennial basis, in order to protect public health and welfare, enhance the quality of water, and serve the CWA's purposes. When adopting or revising standards, states must consider the "propagation of fish and wildlife" and adopt an antidegradation policy to maintain existing instream water uses and the level of water necessary to protect the existing uses. Under section 303(c)(3), if the EPA Administrator determines that a standard meets the requirements of the Act, "such standard shall thereafter be the water quality standard for the applicable waters of that State." EPA's approval of new or revised water quality standards triggers ESA section 7 consultation, which in turn requires that state waters protect endangered and threatened species.

The State of Oregon, for example, has water quality standards for the Willamette Basin that allow "no measurable surface water temperature increase resulting from anthropogenic activities... [i]n stream segments containing federally listed Threatened and Endangered species if the increase would impair the biological integrity of the Threatened and Endangered population." Other designated beneficial uses include anadromous fish passage, salmonid fish rearing, salmonid fish spawning, resident fish, and aquatic life. The state has also adopted narrative standards that are specific to aquatic life criteria. Given such designated beneficial uses and narrative criteria, the state cannot legally issue NPDES permits unless they do not harm listed species, regardless of ESA requirements.

EPA's Oversight Authority of the NPDES and WQS Programs : State assumption of NPDES programs minimizes federal regulatory involvement in water pollution permitting, but EPA retains oversight authority at both the individual permit and program levels. According to EPA's interpretation, however, its NPDES oversight authority does not trigger ESA consultation because there is no federal activity for purposes of section 7(a)(2). Within ninety days of reviewing proposed individual applications, EPA can object to or veto a state permit.

For example, EPA can object where the permit fails to ensure compliance with state water quality standards, such as state narrative or numeric criteria for water quality. EPA regulations, however, do not specifically authorize the agency to object to state-issued permits on grounds that they fail to protect fish or wildlife or ESA-listed species.

Once EPA has raised an objection to the proposed state permit, the state has thirty days to resubmit the permit with new actions to address agency concerns. If the state fails to satisfy EPA conditions, the agency can issue a federal permit. Once EPA approves a proposed permit application, however, it has limited legal authority to require a state to modify an issued permit. EPA also has the authority to withdraw a state program if it fails to meet national standard requirements and to institute enforcement actions on its own. Such power, however, is exercised only in the most egregious circumstances.

Developing water quality standards is primarily the responsibility of states; EPA's role is to develop recommended scientific guidelines for state WQSs and to oversee state adoption and revision of standards that satisfy CWA provisions. EPA's involvement in reviewing new and revised state water quality standards is a federal action, triggering section 7 consultation procedures under the ESA. The CWA requires EPA to develop water quality criteria based on "the latest scientific knowledge... on the kind and extent of all identifiable effects on... plankton, fish, shellfish, wildlife, [and] plant life... which may be expected from the presence of pollutants in any body of water." States, in turn, establish numeric criteria based on either EPA's recommended aquatic life criteria or other "scientifically defensible methods." EPA has authority to approve (within sixty days) or disapprove (within ninety days) new and revised state water quality standards. In cases where EPA disapproves an existing state standard and recommends changes that would achieve compliance with the CWA, the state must adopt such changes within ninety days. If the state fails to adopt minimum federal requirements, the CWA requires EPA to promulgate federal water quality standards.

The Endangered Species Act : Only one year after Congress addressed national water pollution problems, Congress enacted the Endangered Species Act to conserve endangered and threatened species to the point where protection under the statute is no longer necessary. Recognizing the Act's far-reaching and unparalleled protections for species on the brink of extinction, the Supreme Court in 1978 heralded the ESA as "the most comprehensive legislation for the preservation of endangered species ever enacted by any nation." Unlike previous species legislation, the drafters of the ESA expressly declined to protect imperiled species only "where practicable"; instead, they afforded species conservation the highest of priorities regardless of cost. To accomplish this formidable task, the ESA directs the Secretaries of the Departments of the Interior and Commerce to promulgate regulations listing imperiled species as "endangered" or "threatened" and designating their critical habitat. Congress gave the ESA substantive teeth by providing two main prohibitions against adversely affecting listed species.

Which applies only to federal agencies, licensees, and permittees, requires all federal agencies to affirmatively use their authority to promote conservation and recovery of protected species, and it bans all actions likely to jeopardize the continued existence of federally listed species or to adversely modify or destroy the species' designated critical habitat. To comply with the jeopardy prohibition, federal agencies must also follow procedures, which require agencies to consult with FWS or NMFS before proceeding with proposed "agency actions" to determine whether such actions are likely to jeopardize an endangered or threatened species.

Any action "authorized, funded, or carried out" by a federal agency triggers consultation procedures. The Services' joint implementing regulations for consultation broadly construe the term "action":

> *"Action" means all activities or programs of any kind authorized, funded, or carried out, in whole or in part, by Federal agencies in the United States or upon the high seas. Examples include, but are not limited to:...*

The granting of licenses, contracts, leases, easements, rights-of-way, permits, or grants-in-aid; or (d) actions directly or indirectly causing modifications to the land, water or air.

This statutory duty to consult, however, applies only to federal "actions in which there is discretionary Federal involvement or control." Section 9 of the ESA casts a broader net than by prohibiting any person from taking an endangered species within the United States or its waters. The statute broadly defies the term "take" to include "harass, harm, pursue, hunt, shoot, wound, kill, trap, capture, or collect, or to attempt to engage in any such conduct." Both FWS and NMFS define "harm" to include significant habitat modification or destruction that actually kills or injures wildlife by significantly impairing essential behaviour patterns, including breeding, feeding, or sheltering. The Supreme Court has upheld this regulatory definition as a reasonable interpretation of the statute.

Both section 7 and section 9 prohibitions allow limited actions that result in the incidental taking of listed species. Under section 7, where agency actions might otherwise violate the taking prohibition of section 9, FWS or NMFS can issue an incidental take statement along with a biological opinion so long as the Secretary determines that 1) the taking is incidental to the central purpose of the project, and 2) reasonable and prudent measures will minimize taking and avoid jeopardy. Each incidental take statement must include an estimated numerical impact of the incidental take on the species, reasonable and prudent measures to minimize the impact on the species, and any applicable terms and conditions (including reporting requirements). This regulatory exemption is specifically tailored to meet the particular needs of species potentially affected by such federal action. Thus, when EPA administers a federal CWA program, it should consult with the Services on individual permits that may affect endangered species and obtain incidental take statements, provided that the permits minimize harmful effects on aquatic species. Incidental take statements also extend protection from section 9 take liability to permittees.

Section 10 of the ESA authorizes nonfederal parties to apply for incidental take permits. To obtain a permit under section 10(a), an applicant must include a habitat conservation plan (HCP) that identifies 1) impacts likely to result from the taking; 2) measures to monitor, minimize, and mitigate such impacts; 3) available funding to implement these measures; 4) alternative actions to the taking; and 5) reasons why the applicant is not adopting the proposed alternatives and other measures as the Secretary may require. After opportunity for public comment, the Secretary may issue an incidental take permit if 1) he determines that the proposed taking of an endangered species will be incidental to an otherwise lawful activity; 2) the permit applicant will, "to the maximum extent practicable," minimize and mitigate the effects of the taking; 3) the applicant has ensured adequate funding for implementation of the HCP; and 4) the taking of species "will not appreciably reduce the likelihood of the survival and recovery of the species in the wild."

Under state-authorized NPDES permit programs, an individual permit holder cannot obtain a section 7 incidental take statement, because there is no federal action triggering an ESA consultation. Instead, each permit holder must individually apply for an incidental take permit under section 10 and submit an HCP, a time-consuming and expensive process, even if the permittee had complied with CWA requirements when obtaining the discharge permit. In other words, unlike federally issued CWA discharge permits, a state-issued CWA permit is not a shield to an ESA take.

The Intersection of the CWA and the ESA

To adequately address the protection of threatened and endangered species, the CWA and the ESA must work together more effectively. The current statutory framework of these environmental statutes, however, provides only limited overlap given the prominent role of states under the Clean Water Act. In 1973 Congress never addressed whether ESA procedures apply to state-authorized CWA programs. With this congressional silence as a backdrop, EPA has declared that its oversight authority of state-issued CWA permits is not a federal action as contemplated by the ESA and therefore does not trigger the ESA's requirements. Under this interpretation, there is no federal action after EPA approves a state permit program because state CWA programs function "in lieu of" the federal program. EPA, however, does interpret section 7 to apply when 1) it initially approves a state NPDES permit program or 2) when it triennially reviews state water quality standards under section 303, because these approvals are "federal actions" requiring discretion under section 7(a)(2) of the ESA. Section 7 consultation with the Services could establish legally binding water quality standards to protect fish and wildlife. State NPDES permits, in turn, would have to comply with these state water quality standards.

Relying on water quality standards, however, raises important questions about the capability of water quality criteria to support listed species. First, water quality criteria and standards may not necessarily consider all the cumulative effects of pollution if a state has not designated the use of a particular water body for the protection of threatened or endangered species.

Second, even if a state has established the protection of species as a designated use, the water quality criteria may not adequately address all adverse effects on the water quality, such as sedimentation. Finally, there is real concern that agencies would not effectively implement and administer new protective standards given the historical lack of agency enforcement of water quality standards on both the state and federal levels. One recent national study, for example, indicated that approximately thirty percent of all assessed river miles are in nonattainment of fully designated water quality standards. Washington and Oregon both currently violate their water quality standards for temperature in the lower Snake River.

While water quality standards may not produce immediate results in controlling point source discharges, ESA consultations on these standards have the potential to fundamentally change NPDES permit requirements and nonpoint source pollution. If taken seriously, these consultations could compel states and EPA to establish biologically sensitive criteria that protect endangered and threatened species. Moreover, they could lead to more TMDLs and other controls to manage and account for the effects of nonpoint source pollution—potentially the biggest water quality threat to listed species in the West.

EPA's Efforts to Integrate Endangered Species Protections into State CWA Programmes

With limited means to apply ESA consultation procedures to state CWA programs and growing concern about ESA liability for states and individual permittees, EPA, NMFS, and FWS have spent nearly a decade developing an interagency memorandum of agreement to protect endangered species upon state assumption of CWA programs. An initial draft MOA focused on addressing endangered species concerns in water quality standards on a national level. In the subsequent 1997 and 1999 draft MOAs, the agencies expanded their programmatic reach, citing CWA authority to establish federal fish and wildlife agency review of state NPDES permits. This Part reviews each proposed MOA, revealing how the agencies intend to incorporate endangered species protection into state-administered CWA programs.

The 1992 Draft MOA Between EPA and the Services

In 1992 EPA, in conjunction with FWS and NMFS, developed national coordination procedures to protect endangered species upon state assumption of CWA programs. Initially, these federal agencies focused on addressing endangered species concerns in water quality standards. This effort was largely prompted by a lawsuit alleging that the Alabama Department of Environmental Management had violated the ESA by failing to consult with FWS on its water quality standards. In a memorandum of agreement (the 1992 Draft MOA) signed on July 27, 1992, EPA agreed to conduct a national section 7 consultation with the Services for existing national water quality criteria under section 304(a) to ensure that aquatic life criteria for listed toxins adequately protect endangered and threatened species.

This consultation on national water quality criteria would also allow EPA to avoid conducting individual ESA consultations when reviewing state water quality standards that provide identical or more stringent aquatic life criteria than EPA's recommended section 304(a) criteria. However, if the state water quality standard was less stringent than EPA's section 304(a) aquatic life criteria, or if the state set designated uses that failed to provide for the protection and propagation of fish and shellfish, EPA would initiate a separate ESA consultation under its triennial CWA review authority. The 1992 Draft MOA also would establish procedures "to insure that the Services [were] integrated early into EPA's oversight of the states' standards adoption process so that threatened and endangered species concerns [could] be addressed in the most efficient manner possible." Interestingly, while this agreement did not impose consultation obligations on states, it did identify the importance of state water quality standards as an enforceable requirement of all NPDES permits. One article correctly predicted that this MOA would usher in a "practice of initiating section 7 consultation with FWS whenever EPA conducted future triennial reviews of any state's water quality standards." It also suggested that the agencies would apply similar consultation procedures to NPDES permits even though they were "clearly not required by the ESA."

The 1997 Draft MOA Between EPA and the Services

Over the next five years, EPA and the Services attempted to develop more comprehensive coordination procedures to protect endangered and threatened aquatic species during and after state assumption of CWA programs. In July 1997 the agencies proposed a new MOA establishing a national framework for effective.species protection under all CWA programs, including both state water quality standards and NPDES permit programs subject to EPA oversight (1997 Draft MOA). The primary purpose of this draft MOA was to ensure "a uniform level of protection for species regardless of their location, and provide greater certainty for States and Tribes considering assuming a permitting program in the future."

Under the 1997 Draft MOA, the agencies proposed to engage in a "national programmatic consultation" examining EPA's actions when 1) reviewing state water quality standards and 2) approving state or tribal permit programs. This broad national consultation, in turn, would improve existing procedures under the water quality standards program, clarifying how and when EPA must initiate separate section 7 consultations on state WQSs in order to protect listed species. The consultation would also establish coordination procedures between EPA, the Services, and the states under state permit programs to ensure that state-issued discharges adequately protected endangered and threatened species. Most importantly, these procedures were designed to improve cooperation between EPA and the Services and to enhance overall species protection. To integrate endangered species protection into the water quality standards programs, the 1997 Draft MOA first proposed a national EPA rulemaking to amend its national water quality standards and its NPDES permit regulations so that these CWA programs would ensure the protection of federally listed species. The amended WQS regulation, for example, would require EPA to ensure that all water quality standards were not "likely to jeopardize the continued existence of endangered or threatened species or result in the destruction or adverse modification of designated critical habitat."

The proposed rule would also require state policies authorizing variances and mixing zones to avoid jeopardy, as well as the adoption of site-specific water quality criteria protective of listed species. The 1997 Draft MOA also promised that EPA and the Services would work together to develop and review CWA aquatic life criteria. These recommended criteria would establish standards protective of endangered species that states could adopt.

Last and most important, the agreement promised that EPA would institute section 7 consultation procedures when approving new or revised state WQSs. The primary goal of the agreement was to ensure early involvement before and during state development of the standards. To this end, the agreement provided that EPA and the Services would meet to discuss the scope of upcoming state triennial reviews. Where existing standards were not protective of listed species, EPA and the Services would work with the state to obtain revisions to the standards during this triennial review process. The Services' involvement ultimately would help produce water quality standards that protect aquatic species likely to be harmed by permit discharges. Moreover, the repetitious and iterative nature of triennial consultations would allow the Services, EPA, and the states to periodically redetermine the criteria necessary to protect listed or at-risk species.

In the absence of a similar federal nexus to compel ESA consultations on individual state-issued CWA permits, the 1997 Draft MOA identified EPA's approval authority of entire state NPDES programs as the trigger for a programmatic consultation under section 7 of the ESA. Under the agreement, EPA and the Services would conduct a national section 7 consultation to determine whether the proposed coordination procedures for new and existing state permit programs would adequately address species protection. The MOA envisioned a triangular coordination procedure between EPA, the Services, and the states, with EPA acting as the ultimate CWA authority to ensure that state-issued permits address endangered species' needs. The MOA specified that EPA would ensure that states submit all individual permit applications to FWS and NMFS. The Services would then provide the states with a list of aquatic federally listed species and any designated critical habitat in the states. If the Services or EPA believed that an individual permit was likely to have an adverse impact on a listed species, they would contact the state agency to discuss concerns within ten days of receiving the proposed permit.

If the state failed to address species concerns, the Services would contact EPA "not later than five working days prior to the close of the public comment period on the State's or Tribe's draft permit or notice of permit application." EPA would then coordinate with the Services and the state to ensure that the permit complied with CWA requirements, including WQSs, and would discuss "appropriate measures protective of Federally listed species and critical habitat." If a state permit would be likely to have an adverse impact on listed species or critical habitat, EPA could make a formal objection under its CWA authority. In cases where the adverse effects were "minor," the associated take with the individual permit would "be authorized by the incidental take statement accompanying the national programmatic Biological Opinion."

If a state had previously issued a permit that the Services later identified as possibly having an adverse effect on listed aquatic species and that the national incidental take statement did not authorize, the Services would contact the state to remedy the situation. At this stage, however, EPA would have no authority to demand changes regarding species protection to an already issued state or tribal NPDES, sewage sludge, or wetland section 404 permit because of the permit shield. The Services also drafted an accompanying draft programmatic biological opinion to authorize the 1997 Draft MOA's proposed national consultation. Provided that EPA complied with the draft MOA's coordination procedures with the Services, the biological opinion would authorize the incidental take of species whenever EPA reviewed state-issued permits or state water quality standards.

The national incidental take statement, in turn, would provide coverage to dischargers, states, and tribes. However, because the Services could not determine the level of take with numerical precision, they assigned an "unquantified" level of take to this proposed action. Conversely, while the acceptable level of take was classified as minor, neither the draft biological opinion nor the draft MOA itself provided any indication as to what type of point source discharge would constitute a minor impact on aquatic species. Moreover, the explanation of exactly how the Services would anticipate the level of take on a national scale was nebulous at best.

The 1999 Draft MOA Between EPA and the Services

After the circulation of the 1997 Draft MOA, the agencies failed to formally adopt the coordination procedures. In spite of this gridlock, EPA test-piloted the coordination procedures as part of its three most recent federal authorizations of state NPDES permit programs. In Louisiana and Oklahoma, EPA faced judicial challenges from industry stakeholders who argued that the agency had exceeded its statutory authority by instituting federal fish and wildlife agency review as part of the state program approval.

In American Forest and Paper Ass'n v. Environmental Protection Agency (American Forest & Paper I), the Fifth Circuit concluded that EPA could not modify the "non-discretionary" list of enumerated criteria to include review of species when approving a state permit program under the CWA because this list of criteria did not mention protection of endangered species and the ESA conferred no additional substantive powers beyond the underlying statutory authority granted to federal agencies.

The immediate effect of this decision was to vacate the portion of the Louisiana role that imposed the federal fish and wildlife review requirement and authorized EPA to veto any permit to which FWS or NMFS objected. In the second judicial challenge, initiated again by the American Forest and Paper Association, the Tenth Circuit dismissed the case for lack of constitutional standing and ripeness without ever reaching the merits of the Association's challenge. With no clear resolution among these circuits, EPA addressed species protection in its most recent assumption decision, in which it authorized the Texas program, by emphasizing its CWA authority. Specifically, EPA could review state-issued permits to ensure that they complied with state water quality standards protected of listed species.

On January 15, 1999, EPA and the Services jointly published a revised draft of the national coordination program for general public comment in the Federal Register (1999 Draft MOA). Its purpose was to clarify the intent of the interagency agreement, to address comments received from major stakeholders on the previous 1997 draft, and to respond to recent judicial criticism. In the preamble, the agencies expressed the need to integrate the goals of the CWA and ESA on a program-wide level, thereby ensuring "an appropriate level of protection for listed species and greater regulatory predictability for States, Tribes, and the public."

The interagency procedures would serve agency goals by achieving greater efficiency and avoiding protracted consultations, such as separate state WQS consultations that take an average of eighteen months each. In advancing this program, EPA dismissed the significance of the outcome in the Fifth Circuit case from Louisiana, asserting that the agency "believed that this case was wrongly decided" because the court ignored the agency's discretion in reviewing an entire state permit program under the CWA. Moreover, EPA would retain final authority in evaluating whether a proposed discharge would harm a listed species pursuant to its mandate under the CWA and not the ESA.

Although the 1999 Draft MOA continued to focus on incorporating species protection into the water quality standards and NPDES permit programs, it presented a number of important changes from the 1997 Draft MOA.

The agreement, for example, attempted to streamline individual consultations on state water quality standards by simply having states adopt federal aquatic life criteria already approved by the Services in the MOA's national consultation. Moreover, the 1999 Draft MOA clarified the existing authority under the CWA to implement the proposed interagency procedures under state permit programs. Overall, in light of the cooperative federalist nature of the CWA, the 1999 Draft MOA represented an important effort to integrate and further the goal of sustainable aquatic ecosystems.

Instead of amending both the WQS and NPDES permitting regulations, the 1999 Draft MOA limited its proposed national rulemaking on species protection to the water quality standards regulations. EPA asserted that the rule will essentially "codify existing protection for endangered and threatened species under the CWA." In EPA's judgment, water of such degraded quality "that it will likely cause jeopardy to the continued existence of a species would generally not be consistent with protections provided by the Clean Water Act." More significantly, the 1999 Draft MOA proposed to conduct a national section 7 consultation on EPA's recommended aquatic life criteria to determine the effect of the criteria on listed species and designated habitat. States, in turn, could adopt EPA's water quality criteria as part of their water quality standards, thereby avoiding a separate consultation on such adopted criteria.

The agencies claimed that, in contrast to case-by-case state consultations on water quality standards, a national consultation on aquatic life criteria would promote greater efficiency and ensure agency consistency in assessing the impacts of pollutants on aquatic endangered species and in developing protective standards. EPA pointed to recent success stories in which either the agency or the state included aquatic life criteria in WQSs based on EPA's section 7 consultations and the accompanying biological opinions. The MOA's national consultation would provide incidental take coverage for any state water quality criteria that are at least as stringent as EPA's recommended section 304(a) criteria.

Under the state permit program, the 1999 Draft MOA justified all coordination procedures with the Services under existing Clean Water Act authority. This statutory emphasis appeared to respond to the Fifth Circuit's criticisms that EPA was limited to working within its existing CWA authority to address the needs of aquatic species. For example, EPA explained that it would approve a state permit program based only on the criteria enumerated in section 402(b) of the CWA, and that it would not place any other conditions on agency approval to protect endangered species. As part of its oversight authority, however, EPA would object to individual state-issued permits likely to jeopardize listed species on the grounds that such permits are "outside the guidelines and requirements" of section 402(d)(2) of the CWA. Further, in contrast to the 1997 Draft MOA, which required states to submit permit applications to the Services, the 1999 coordination procedures clearly identified existing CWA regulations that require states to provide notice and copies of individual draft permits to the Services. By grounding the procedures in current regulatory requirements, EPA thus bolstered its argument that these interagency efforts to protect ESA-listed species in CWA programs do not impose any obligations not already required by the CWA. Last and most important, the 1999 Draft MOA referred back to the 1992 draft interagency agreement and identified water quality standards as the key to ensuring protection of endangered and threatened species. The document also reiterated similar portions of the 1997 Draft MOA, identifying the crucial link between state permits and water quality standards:

Water quality criteria and State or Tribal water quality standards establish levels of pollutants from all sources, and so would account for all such effects insofar as water quality is concerned. Since NPDES permits are established to achieve water quality standards, they will account for point source effects insofar as water quality is concerned.

The 1999 Draft MOA went further by explaining that water quality standards enable EPA to consider the effect of permit discharges on endangered species under its oversight authority. At the same time, however, the Services have openly admitted in recent biological opinions that many state water quality standards currently cannot afford proper protection to endangered and threatened species. In the biological opinion approving Texas's state permit program, for instance, FWS wrote that "research indicates that species in or near contaminated sediments may be adversely affected even if water-quality criteria are not exceeded." Moreover, the current Texas state WQS for temperature cannot "adequately protect reproduction and survival of the fountain darter." Many other current state water quality standards also may not be adequate to protect listed species.

Evaluating EPA's Current Approach to Species Protection

The 1999 Draft MOA stated that "EPA and the Services believe that this approach ensures that any issues regarding listed species will be adequately addressed." It may be too early to determine the full extent of EPA's approach to integrating species protections into state CWA programs, but an assessment of the effectiveness of EPA's efforts thus far is possible. Successful implementation of EPA's programs will require state permit programs to rely on water quality standards protective of listed species and consultations of water quality standards that demand water quality goals for aquatic viability.

With these overarching objectives in mind, this Part examines the 1999 Draft MOA, commending the agencies' efforts to achieve species protection via water quality standards, while questioning the legality and biological integrity of conducting a national section 7 consultation. It then examines the potential for water quality standards to meet the formidable burden of advancing endangered species protection in both the approval of Texas's permit program and the triennial review of Oregon's water quality standards.

Evaluating the 1999 Draft MOA

A review of the 1999 Draft MOA clearly indicates a concerted and innovative effort to integrate species protection into state-administered water quality programs. In essence, the ESA is driving water quality standards to play a key role in protecting the needs of endangered and threatened species under the CWA. Accordingly, because EPA's triennial review of state WQSs triggers section 7 consultation, states would promulgate water quality standards that create viable habitats for listed species. EPA, in turn, would object to state-authorized permits that violate WQSs under its CWA oversight authority. In other words, such a permit would be "outside the guidelines and requirements" of the CWA. Such efforts to work within the existing CWA regulatory scheme should prevent judicial reversals like that in the Fifth Circuit.

Despite the 1999 Draft MOA's commitment to species protection, it faces some formidable challenges. First and foremost, it is unclear how the draft MOA's national formal section 7 consultation on water quality criteria and standards and its coordination procedures will satisfy the ESA's biological opinion and incidental take statement requirements. The 1999 Draft MOA asserts that the Services' accompanying biological opinion will authorize the incidental take associated with state-issued NPDES permits, which incorporate national state water quality criteria, and provide coverage to dischargers, states, and tribes so long as the coordination procedures are followed. The biological opinion also will authorize incidental take of any state water quality criteria consistent with or more stringent than the recommended national criteria. Given the scope of this national biological opinion, however, the 1999 Draft MOA admits that it cannot accurately quantify the level of "take" or consistently identify the instances of adverse effects:

The consultation will be conducted on a national basis, and therefore, will not be water-body specific. In addition, given the numbers of species involved in the consultation, the effects on species will be evaluated to the maximum extent possible based on groupings of species believed to be affected in a similar manner. As currently envisioned, this programmatic incidental take statement would violate section 7 procedure, s that require biological opinions to 1) specify the effect of incidental take on endangered or threatened species, 2) establish monitoring requirements and reasonable and prudent measures to minimize incidental take, and 3) create terms and conditions that identify a numeric take level if possible or at least a descriptive one. Without an evaluation of site-specific information and effects on particular listed species, the draft MOA inadequately provides for species survival and recovery. Moreover, given the inherent ambiguity and complexity of programmatic consultation, it is less likely that EPA will reinitiate consultation under the ESA regulations, which trigger consultation once "new information reveals effects of the action that may affect listed species or critical habitat in a manner or to an extent not previously considered."

Second, as a federal agency, EPA has affirmative duties to conserve listed species under ESA section 7(a)(1) and to consult with the Services on how the agency's actions might affect each of the listed species; EPA cannot "just undertake a generalized consultation" as the 1999 Draft MOA proposes. In order to develop species-specific conservation programs and comply with section 7(a)(1) obligations, EPA should evaluate listed species by water body and ecosystem for both individual water quality standards and permiting actions.

Additionally, in assessing species believed to be similarly affected, the Services would contravene the ESA's requirement to use the best available scientific information. Thus, in an attempt to streamline the ESA process with a sweeping incidental take statement covering individual state-issued permits and water quality standards, the 1999 Draft MOA fails to satisfy the ESA. Before the agencies issue the final MOA, they should reconsider whether a national consultation is actually capable of providing the level of protection required under the ESA, and if so, they should establish clear standards in developing a national consultation.

Third, even though EPA has discretion to decide whether point source discharges will adversely affect listed aquatic species, the agency is constrained by the CWA, which provides no mechanism to require changes once a state has already issued an Individual permit that is adversely affecting federally listed species. In such an event, the 1999 Draft MOA proposes that EPA and the Services request that the state reopen the permit. This option is unsatisfactory, given that one of the purposes behind the coordination procedures is to avoid potential jeopardy problems for state-issued permits in the first place. Moreover, the Services could potentially sue the permittee under ESA section 9 for the resulting take, because the 1999 Draft MOA and biological opinion would authorize incidental take of species only where EPA had properly followed the coordination procedures with the Services.

Fourth, the coordination procedures for the permitting program may not provide sufficient time for the Services to adequately compile information and make an accurate determination about whether a state-authorized discharge will jeopardize aquatic species in the relevant waters. The 1999 Draft MOA requires the Services to express "concerns" to the appropriate state agency, "preferably within ten days of receipt" of a proposed permit application. If the Services fail to initially address this concern with the state, the agencies' last opportunity is to contact EPA directly not later than five days prior to the dose of the permit application's public comment period.

Given the broad scope of the Services' duties, a ten-day period followed by another fifty-five-day period may be too short of a time frame. In addition, EPA can choose to ignore the Services' jeopardy concern under the coordination procedures more readily than it could a jeopardy determination in a biological opinion issued under formal ESA consultation procedures.

Because biological opinions issued pursuant to the ESA receive considerable deference from reviewing courts, EPA can rarely ignore an ESA biological opinion without running a significant risk of violating section 7. In contrast, under the proposed agency scheme, a reviewing court is likely to defer to EPA's judgment given that there are CWA procedures and a sparse biological record for the court to review. To remedy these problems, the MOA should extend the period for the Services to review potentially harmful individual permits, thereby allowing the Services to develop a more precise and detailed record.

Fifth, from the perspective of the permit holder, the agency coordination procedures yield mixed results. Permit holders may feel a certain level of comfort with the agencies' national incidental take statement, because it seems to shield individual permittees from the taking prohibition of ESA section 9. At the same time, some permittees may oppose interagency procedures, claiming that the additional mandatory review will lengthen permitting delays, increase costs, and generate burdensome permit provisions. If EPA, for example, decides to object to a proposed state permit because of endangered species concerns, the applicant might have to obtain a separate incidental take permit under section 10—a time-consuming and expensive process. In addition, some industry representatives have attacked the proposed national rulemaking on WQSs as imposing "new" water quality requirements to protect listed species. This rule, they contend, will result in the banning of mixing zones in certain locations, thereby preventing dischargers from temporarily exceeding water quality standards at the end of the pipe.

Finally, some environmentalists may view these interagency procedures as better than the status quo of existing state approved programs. In contrast to traditional oversight authority between EPA and the states, the 1999 Draft MOA would create federal fish and wildlife review of state-issued NPDES permits and water quality standards protective of listed species. The Services' participation in reviewing state NPDES permits and WQSs theoretically should reduce the likelihood of harm to listed species in state waters. Some commentators, however, argue that EPA could easily call on its oversight authority to trigger traditional section 7 formal consultation procedures when reviewing new and renewing old state-issued permits. Arguably, EPA has discretion comparable to its state approval authority, which would appear to satisfy the discretionary element of the ESA regulations. The legal case for subjecting individual state-issued NPDES permits to ESA consultations may be weak, however, because courts have held that EPA's authority to veto state permits does not involve a federal action.

Approval of the Texas NPDES Permit Programme

On September 14, 1998, only thirteen days after the Tenth Circuit's opinion dismissing the challenge to Oklahoma's state program, EPA approved Texas's application for the administration of the NPDES permit program by the Texas Natural Resource Conservation Commission (TNRCC). The MOA between EPA and TNRCC established virtually identical consultation procedures as established by the agreements in Louisiana and Oklahoma, providing opportunities for 1) the Services to comment on state permits, 2) the Services to contact EPA if their comments were not adequately addressed by the state, and 3) EPA to object to state NPDES permits if they fail to meet the requirements of the CWA.

Given a negative precedent in the Fifth Circuit, however, EPA went to great lengths in its rule approving the Texas program and its MOA with TNRCC to distinguish it from both its Louisiana and Oklahoma approvals. This distinction was crucial because Texas is in the same circuit as Louisiana. The EPA rule explained that the agency would "only object to a State permit where doing so would be within its authority under the CWA." EPA also directly addressed the Fifth Circuit's holding that EPA had exceeded its statutory authority by conditioning state approval on state compliance with the ESA. The agency emphatically pronounced that it had "not `conditioned' its approval of Texas's application on any factors related to endangered species protection." Instead the agency claimed that it had approved Texas's program "based solely on the criteria contained in section 402(b) of the CWA and implementing regulations."

The most compelling agency argument in support of its CWA authority involved enforcing state water quality standards protective of listed species as part of state-issued NPDES permits. In contrast to the Louisiana and Oklahoma MOAs, the Texas MOA directed the state to "address effects on endangered species and historic properties through setting and enforcing water quality standards." EPA relied on section 301(b)(1)(C) of the CWA and its implementing regulations, which require NPDES permits to comply with more stringent limitations, including state WQSs, than mere technology-based effluent limitations. As part of its CWA oversight authority, EPA thus could object to a state-issued permit if the discharge failed to meet state WQSs designed to protect listed species. In Texas's case, the state WQSs also specifically included a narrative standard provision protective of aquatic species: "Water in the state shall be maintained to preclude adverse toxic effects on aquatic and terrestrial wildlife... resulting from contact, consumption of aquatic organisms, consumption of water, or any combination of the three." Under this scheme, EPA clearly articulated the important connection between water quality standards and NPDES permits and revealed how these two interrelated CWA mechanisms could work in tandem to serve the needs of listed aquatic species.

As part of a formal section 7 consultation on Texas's assumption, FWS concluded in its biological opinion that the triangular coordination procedures involving the Services were adequate to ensure that state-issued permits were not likely to jeopardize listed species and would minimize incidental take. Accordingly, FWS issued a sweeping incidental take statement as part of federal authorization of the Texas NPDES (TPDES) program. This incidental take statement authorized "minor" takes associated with TPDES permits, unless the Services submit a written concern to EPA about potential adverse effects a proposed TPDES permit might have on a federally listed species or critical habitat. Under this scheme, the burden rested on the Services to identify and raise a jeopardy concern in a proposed TPDES permit within specified time limits. Otherwise, all state-issued permittees and state authorities were exempt from liability under the section 9 taking prohibition of the ESA.

Texas's biological opinion raises similar problems to the 1997 and 1999 Draft MOAs. For example, while FWS loosely defined a "minor" take, it avoided numerically identifying the extent of the take within the waters in the State of Texas. Instead, FWS acknowledged the difficulty of quantifying the amount of take in the entire state given that the take associated with "the future issuance of TPDES discharges for as yet unknown discharges potentially affect[s] all geographic areas." Interestingly, FWS indicated some reservation about relying on current Texas WQSs in state-issued permits as a means of protecting endangered species. The Service stated that "implementation of those [Texas Surface Water Quality] standards through permitting has not been (nor were they intended to be completely) protective of individuals of a particular species." Specifically, FWS anticipated take in the form of harm in mixing zones and downstream of permittees' outfalls. In fact, the agency conceded that "take in the form of harm or harassment may be occurring presently as a result of EPA's administration of the NPDES program," thus revealing an urgent need for immediate consultation on this federal agency action. Most troubling of all was the fact that this approved state program rested on water quality standards for which EPA had yet to complete a section 7 consultation.

The biological opinion for Texas reveals that there will be considerable lag time between approval of the state permit program and actual protection of listed species under the CWA authority because 1) many current water quality standards—the baseline for NPDES permits—are harmful to threatened and endangered species, and 2) significant changes under the permitting program will be incremental, given that a state cannot impose new effluent limitations during the permit's five-year lifespan. To address these pragmatic considerations, EPA should expedite consultation on state water quality standards and authorize incidental take coverage to the states and individual permittees pending approval of water quality standards that are in fact protective of threatened and endangered aquatic species.

Section 7 Consultation on Oregon's Water Quality Standards

In July 1999, three and one-half years after the Oregon Department of Environmental Quality (ODEQ) submitted its water quality standards to EPA for review, NMFS finally issued a biological opinion. This represented the first time in the region that a section 7 consultation examined a state's water quality standards. The purpose of the biological opinion was to evaluate whether EPA's proposed approval of Oregon's water quality standards for dissolved oxygen, temperature, and hydrogen ion concentration (pH) standards would likely jeopardize the continued existence of any of the proposed and listed salmonid species.

In 1996, when consulting on the triennial review of Oregon's standards, EPA determined that ODEQ had failed to implement and enforce the existing temperature standards and expressed concern that the standards were not adequate to protect listed salmonid species, The biological assessment, in fact, determined that salmon were "likely to be adversely affect[ed]" by the standards, which "pose[d] a risk to their viability." EPA nevertheless proposed to condition its approval of Oregon's standards on several "conservation measures"—including a regional review of temperature standards over the next two years—that the state would consider adopting within the next triennial standards review.

The belated NMFS biological opinion on Oregon's failing water quality standards proved a great disappointment. Analysing the state's dissolved oxygen standards, NMFS concluded that the cool water standards were "likely to adversely affect" threatened salmon species in Oregon and Washington because the state had established many standards based on weekly rather than daily averages. As for temperature standards, NMFS agreed with EPA's determination that a sixty-four degrees Farenheit temperature was "likely to cause lethal and sublethal adverse effects," including increased mortality and disease for chinook. Moreover, NMFS determined that the spawning temperature criterion of fifty-five degrees Farenheit was inadequate given the state's inaccurate spawning dates table. Nevertheless, NMFS issued a conditional no jeopardy opinion, even though the agency concluded that "any further degradation of these conditions would significantly reduce the likelihood of survival and recovery of these species due to the amount of risk that salmon face under the current environmental baseline."

These conditions included EPA's earlier "conservation measures," such as a regional review of temperature standards over the next two years and adoption by Oregon of these or similar standards. To satisfy section 7(a)(1)'s duty to conserve, the biological opinion provided several conservation recommendations: several research studies to analyse historic water temperatures in ecoregions of the state; models of attainable stream temperatures based on changes in vegetation, flows, and restored hydrologic connections to groundwater and floodplains; and fish studies on how sublethal temperature levels affect salmon reproduction and distribution patterns. In addition, NMFS agreed to reassess the conservation measures implemented by EPA and Oregon every six months to determine if consultation must be reinitiated.

Despite these promised conservation and mitigation efforts, this biological opinion concludes that Oregon's existing water quality standards for dissolved oxygen and temperature do not protect all threatened and endangered species. The regional review of temperature standards, for example, fails to adequately ensure that temperature standards will be more protective than the current standard because of the extended timeframe between the review and actual adoption of the new or equivalent standards by the State of Oregon. This waiting period to implement such mitigation measures clearly violates the ESA's mandate, as it postpones confronting the adverse effects of water quality standards on listed species. Some commentators further lament that "[t]he likelihood that the state, or EPA subsequently, will take actions consistent with their own science is remote" because of increased political pressures from industry. Moreover, the biological opinion on Oregon WQSs ignores one of the requirements of the consultation, which demands efforts "necessary for the listed species to survive and recover to naturally reproducing population levels" so that "protection under the ESA would become unnecessary." In short, Oregon's biological opinion and recommendations fail to achieve the purpose of section 7 consultation—securing immediate protection for species facing extinction.

Recognizing the vital connection between species and their habitats, the ESA calls for "a means whereby the ecosystems upon which endangered species and threatened species depend may be conserved." To meet this objective of ecosystem conservation, as well as the CWA's fishable water quality goal, the ESA and CWA must function together so that water quality is capable of protecting listed species. In an attempt to ensure heightened species protection and provide greater assurance against liability to state authorities and their respective permittees under the CWA, EPA proposes to institutionalize section 7 consultations for water quality standards and to develop federal fish and wildlife review of state permit programs. This effort represents an important shift in agency thinking, viewing water quality standards as the key to trigger endangered species protection in state-administered CWA programs. Thus far, EPA has faced two judicial challenges questioning the scope of the agency's authority to mandate these new procedures under permit programs. With only one case decided on its merits, the new agency framework has not been fully tested.

More importantly, though, it is unclear when water quality standards will actually protect species and, in turn, serve as the environmental baseline for state-issued permits and nonpoint source pollution. Currently, Texas's permit program, which incorporates the draft MOA's coordination procedures, is enforcing water quality standards that harm listed species. While section 7 consultations on water quality standards could fundamentally alter state CWA programs, consultations like Oregon's are far from revitalizing the standards, thus falling short of the level of protection required to return endangered species to the point where they no longer need this crucial protection. At the very least, though, Oregon's water quality standards will trigger another ESA consultation every three years, compelling EPA, the Services, and the state to reassess and determine how they can achieve fishable habitat. In the end, if water quality standards are to be the mechanism to protect listed species under the CWA and complement the goals of the ESA, as EPA and the Services advance, consultation on water quality standards must afford actual protection for listed species, and state permit programs must implement these protective water quality standards.

Neighbour's Water: Solving the Water Regulation

When you think Texas, you likely picture the current President of the United States of America, George W. Bush. And, when you think President Bush, you likely think of the war in the Middle East. Regardless of political views on the matter, this serves as a connection between Texas and the Middle East. Yet, there is another, less obvious, connection that exists between these two areas: both face the challenge of how to structure their water regulation systems in order to preserve water for future generations. After looking at the insufficient water regulations of Texas and countries in the Middle East, it becomes clear that both areas will need to undergo a massive overhaul in water regulations to ensure they preserve access to water. Each area will present two separate problems with regard to water regulations, but both may benefit from the same proscriptive measures in attempting to solve their water regulation problems.

Original Adoption

The English rule of capture was adopted in the 1904 Texas case of Houston & Texas Central Railway Co. v. East. In East, the plaintiff brought suit against the defendant railroad company for damages caused by digging a well that made the plaintiffs well run dry. The Texas Supreme Court relied on the principles set forth in the English case of Acton v. Bundell to determine that one had a right to dig a well and, if that digging resulted in the drying of a neighbour's well, it would be deemed unactionable. The East court necessitated the rule of capture doctrine by looking toward the principles set forth in Frazier v. Brown. In Frazier, the Ohio Supreme Court found that no correlative rights with regard to underground percolating waters were recognized by law. In justifying its decision, the Ohio court focused partly on policy considerations:

Because the existence, origin, movement and course of such waters, and the causes which govern and direct their movements, are so secret, occult and concealed, that an attempt to administer any set of legal rules in respect to them would be involved in hopeless uncertainty, and would be, therefore, practically impossible. 2. Because any such recognition of correlative rights, would interfere, to the material detriment of the common wealth, with drainage and agriculture, mining, the construction of highways and railroads, with sanitary regulations, building and the general progress of improvement in works of embellishment and utility.

During this time, courts treated the rule of capture as a tort doctrine with regard to neighbours and the pumping of water.

From Tort Doctrine to Vested Property Right

In Texas Co. v. Burkett, the Texas Supreme Court fortified the rule of capture by interpreting it to be a vested property right and no longer a tort doctrine. In Burkett, the plaintiff sought to enact a contractual agreement with the defendant. The defendant argued that the plaintiff did not own water and that the water actually belonged to the state; therefore, the plaintiff did not have a right to sell. The court disagreed with the defendant and ruled that the plaintiff "plainly had the right to grant access to [streams] and the use of their waters for any purpose...." The court in Burkett created a common law rule classifying water ownership as a vested property right. There has been criticism with regard to groundwater falling under the umbrella of property rights. Water ownership cannot satisfy the characteristics of an "efficient property rights system" including universality, exclusivity, transferability, and enforceability.

Defining Absolute Ownership

After Burkett, other Texas courts began following the lead by interpreting East as expressing an absolute ownership principle. The Texas Legislature codified this sentiment by enacting the Texas Underground Water Conservation Act of 1949, which recognized ownership and rights in underground water of "the owner of land, his lessees and assigns." The years following the state legislature's adoption of the absolute ownership principle gave preference to protecting the water rights of farmers and ranchers.

This can be largely attributed to the rural influence present in that particular legislature. Not until the 1960s—when legislative seats were being redistricted—did the Texas Legislature witness a rise in urban interests. These urban interests viewed water in the agricultural sense until the 1980s.

Notably, a majority of other states do not recognize private ownership of groundwater. It is important to distinguish two types of water when discussing water ownership: surface water and groundwater. Both surface water and groundwater have "different allocation rules, distinct conflict resolution frameworks, and separate administrative agencies." In recognizing the value of water resources, however, the Texas Legislature has expressed a common goal of effective utilization of both groundwater and surface water sources to meet the present and future water needs in Texas.

Now Enter Senate Bill 1

In 1997, the Texas Legislature enacted Senate Bill 1, which demonstrated that groundwater regulation should be "tailored to meet local water supply and population needs." In order to accomplish this goal, the Texas Legislature created Underground Water Conservation Districts (UWCD). Texas regulates its groundwater with UWCDs instead of using a regulatory scheme that creates a uniform set of rules across the state. Subject to regional and state approval, UWCDs give residents who live close to water sources the power to establish rules with regard to water management. This requires that each district analyse both available and anticipated water supply and demand in that particular district. Once this analysis is completed, districts are to develop a plan addressing their management goals and objectives. After this plan is submitted and approved by the Texas Water Development Board (TWDB), it is then integrated into the regional water plan and serves as the basis of action for a district.

Present Day Ownership

The present day understanding of the rule of capture rests on the idea that a "landowner possesses the water beneath the surface by virtue of control over the surface." Three principles guide the rule of capture in Texas: (1) water under the land belongs to the owner; (2) the owner or assignee may pump as much water as he likes without being held accountable to others affected; and (3) a party may sell whatever water is pumped to third parties, and such water may be used in other locations.

To better understand the rule of capture, one must examine the limitations in place for owning, pumping, and selling groundwater. One limitation on water ownership is the defining of exactly what water belongs to the state and what water belongs to the surface owner. There are four categories of groundwater sources that determine to whom the water belongs: (1) underflow, (2) underground stream, (3) artesian, or (4) percolating.

Underflow of a watercourse and underground streams in defined channels belong to the state. Watercourse underflow is defined as the portion of a surface watercourse that flows through sand and gravel deposits beneath the surface of the stream bed. Courts will treat underground streams in defined channels as surface water—thus, not the sole property of the surface owner—if it has the same characteristics of a surface watercourse. These characteristics include the presence of banks, beds that form a channel, and a current of water. The importance of the artesian water restriction involves statutory provisions that prevent waste of this source and require permits for its withdrawal. Percolating water belongs solely to the surface property owner. Once a landowner has ownership over percolating water, further limitations are set in place to restrict what kind of usage may occur.

There are two common law restrictions on a landowner's right to capture, pump, and sell percolating water. First, a surface owner may not "maliciously take water for the sole purpose of injuring his neighbour... or wantonly and willfully waste it." There is, however, criticism of this limitation because some feel that proving the malice intent element is too difficult and, therefore, makes such a limitation ineffectual.

Additionally, there is a belief that the Texas Supreme Court created a major escape hatch by stating that "waste" refers only to "waste in use" and not "water lost during transport." This distinction between use and transport was articulated in City of Corpus Christi v. City of Pleasanton where the court allowed a city to extract ten million gallons of water a day for transport without classifying it as waste. The Texas Supreme Court, however, later came back in Bragg v. Edwards Aquifer Authority with a broader definition of "waste" as it applied to the rule of capture limitation. It has been suggested that Bragg "opened the door for future Texas Legislatures to finally empower locally elected groundwater conservation districts with the powers necessary to regulate groundwater withdrawals without running afoul of the rule of capture."

The second limitation stipulates that a landowner may not cause subsidence in neighbouring land to occur through use of water on his land. This limitation was created in Friendswood Development Co. v. Smith-Southwest Industries where the Texas Supreme Court ruled that a negligent subsidence exception allows for deterrence of excessive underground water withdrawals and resulting land subsidence. The basic policy behind the court's ruling rested on the idea that "ownership and rights of all landowners will be better protected against subsidence if each [landowner] has a duty to produce water from his land in a manner that will not negligently damage or destroy the lands of others."

With these restrictions in place, however effective they may or may not be, the rule of capture still allows a landowner to appropriate waters while on his land and make whatever use of that water he chooses. All of this is without regard to the fact that his use may cut off the flow of water toward neighbouring land, thus depriving that neighbouring landowner of water use. As one might imagine, the rule of capture has led to controversy, but more importantly, it has also led to a massive depletion of water supplies. A major part of this depletion can be accredited to the large growth of urban areas in Texas, which has put a strain on groundwater resources.

Persistence of the Rule of Capture

With all of its negative effects and the inundation of criticism with regard to the rule of capture, why do courts still adhere to it? Two justifications have been posed to help answer this question: reliance on the rule and legislative deference. For courts adhering to the rule of capture, reliance is an ever-weakening argument as time progresses and water management problems increase. One rationale that has been offered to explain why reliance is a diluted basis for following the rule of capture is that although a landowner can rely on this rule to release him from liability for pumping his neighbour's well, this same landowner, in turn, realizes that the rule does not protect his well from his neighbour. In fact, rural parties, such as farmers and ranchers—once the most devoted of supporters for the rule of capture and antiregulation —now stand as parties calling for regulation in order to create "firm water rights."

This leads to the next justification for sticking with the rule of capture: legislative deference. Texas courts point to the Texas constitution, which states that "the preservation and conservation of all such natural resources of the State are each and all hereby declared public rights and duties[,] and the Legislature shall pass all such laws as may appropriate thereto." An example of such application is Sipriano v. Great Spring Waters of America, in which the court stated that "[i]t would be improper for courts to intercede at this time by changing the common law framework within which the Legislature has attempted to craft regulations to meet this state's groundwater-conservation needs." This decision was due to the then-recent passing of Senate Bill 1 and the legislature's creation of a water district system to address water concerns for individual areas.

Although the Texas Supreme Court has yielded to the legislature, lower Texas courts have not given such deference to Senate Bill 1. In South Plains Lamesa Railroad v. High Plains Underground Water Conservation Dist. No.

1, plaintiff landowner brought action against the water district for revoking a water well permit and denying another. The court held that the water district lacked the authority to deny and revoke water well permits for "disproportionate taking" in relation to track size and, by doing so, exceeded their authority. Discussing the Texas Supreme Court's prior decisions, the court in South Plains determined it was in a position to define aspects of the water district's power rather than deferring to the legislature. This discrepancy between the more activist lower courts and the Texas Supreme Court has led some to conclude that although courts defer to the legislature, they are also the ones who are impeding the legislature from changing water regulation.

Middle East Background

The Middle East, as a whole rather than a single country, offers a beneficial comparison to the Texas water regulation system. This is due to the water problem in the Middle East, as in Texas, being a growing problem among neighbours. Water has always been scarce in the Middle East, but the growing need from increased populations and agricultural sectors is elevating this scarcity toward the level of crisis. A major part of this problem arises from the fact that most countries in the Middle East share the same aquifers. This problem is exacerbated by the number of countries that do not take into account their impact on a neighbouring country's water resources. It has been argued that the conflict over water rights serves as a major barrier to reaching a peaceful relationship among the countries in the Middle East.

Setting the Stage for Conflict : Use, share, control, and management of water creates high levels of tension among Middle Eastern neighbours. The Middle East, along with North Africa, represents five percent of the global population, but only.09% of global water resources. A problem arises for the Middle East due to the Tigris, Euphrates, and Jordan River systems serving as the only main waterways for the region. By 1990, there were eleven water-scarce countries in the Middle East and North Africa. By 2025, it is predicted that an additional seven countries in this region will be added to the list.

A number of factors have contributed to the increasing scarcity of water in the region. First, population rates in the Middle East are growing rapidly, and water is being consumed at such a rate that it is impossible for the water to replenish itself naturally. This growth has also caused increased pressure for economic development and irresponsible patterns of urbanization and industrialization. Second, domestic pollution has contributed to the deterioration in the quality of available water. This can partly be attributed to the lack of environmentally conscious technology that has accompanied the industrialization of the Middle East. Finally, the expansion of agriculture and industry has also eroded available water resources in the region. These three factors combined have caused the per capita water availability in the Middle

East to become one of the worst in the world. Because water resources are continually shrinking, a number of Middle Eastern countries have turned to using treated wastewater for industry, agriculture, and the recharging of aquifers. This, of course, gives rise to health risks due to the bacteria, viruses, and organisms present in the water. One of the factors contributing to tension over water in the Middle East is the fact that upwards of fifty percent of the population has to rely on water coming from another sovereign state. This transboundary aspect, along with the scarcity of water in the Middle East, has been the spark of a number of conflicts. Such conflicts include ones among Turkey, Iraq, and Syria over the Tigris and Euphrates River; among Israel, the Palestine Authority, and Jordan over the Jordan River Basin; and within Saudi Arabia itself.

Turkey—Syria—Iraq: Since the 1970s, there has been a growing tension between Turkey, Syria, and Iraq over water. This is due to the Tigris-Euphrates River system being the only real source of water in the region. For six thousand years, the basins formed by both the Tigris and Euphrates Rivers have been part of a number of struggles for domination. Turkey sits toward the north of both Syria and Iraq. For Syria, the main water source is the Euphrates River, which flows down from Turkey and then into Iraq. Iraq, on the other hand, relies both on the Euphrates River water supply as well as water from the Tigris River that flows from Turkey into Iraq, thus giving them an additional source of water. Notably, Iraq is the country with the highest consumption of water from the Euphrates River among the other countries in the region. In 1987, a protocol was signed, which allowed for Syrian access to the Euphrates River Basin. However, "Turkish development efforts have increasingly threatened to marginalize and even eliminate Syrian access to water." One such development effort is the Southeast Anotolia Project (GAP). The goal of this project was the development of the area consisting of nine provinces in the Euphrates-Tigris basins known as the "GAP Region". Originally, the GAP started as an energy production and irrigation project but has since been expanded to cover urban, rural, and agricultural infrastructures. It has been argued that this development has essentially given Turkey control of the only upstream water source for Syria. In fact, the GAP will have a great impact on the water supply of Syria and Iraq flowing from the Euphrates River.

Jordan—Palestine—Israel: Turning toward another water conflict in the Middle East, the Jordan, Palestine, and Israel region demonstrates the problems of overuse due to population growth and pollution. Their water source originates in Lebanon and has a total average flow of around 1,200 cubic meters a year. This water system is made up of both the Jordan and Yarmouk Rivers. The Jordan River basin is viewed as the most controversial in the region due to it being the most developed and politically intricate. Because a number of states that rely on the Jordan River for water are not able to harvest enough water to meet their needs, they must also rely on groundwater aquifers as a main source of water. This reliance is due to Jordan's per capita water consumption, which, although one of the lowest among countries in the Middle East, is still more than the amount of water it can harvest from rainfall and from the Yarmuk River. In 1994, the Israeli-Jordanian Treaty was signed, creating guidelines regarding the distribution, preservation, and availability of water from the Yarmouk and Jordan Rivers. However, despite this agreement, conflict is continually rising. In 1999, the climate of the region and harsh droughts caused Israel to be unable to adhere to the stipulations of the 1994 treaty and cut the annual allocation of water to Jordan by half. This caused Jordan to be unable to sustain its current levels of water consumption and move toward a water rationing system. Because of Jordan's limited access to water resources, its escalating water deficit is expected to reach 250 cubic meters by the year 2010. With Jordan placing a high value on the "hydraulic imperative" and Israel's decreasing internal water sources, fears of a grab-for-resources are increasing among the two countries.

Israel, on the other hand, has relied almost entirely on its National Water System, which is made up of the West Bank Mountain Aquifer, the Coastal Aquifer, and the Lake Kinneret Basin. Because of this reliance, the "Mountain Aquifer underneath the West Bank is a point of contention between Israelis and Palestinians." More specifically, one of the major issues between Israel and Palestine is Israel's blocking of Palestinian access to water supplies. This has caused Israel to overexploit available resources in order to expand agricultural and industrial ventures.

Saudi Arabia: Saudi Arabia serves as another demonstration of a country quickly approaching a water crisis. The difference between Saudi Arabia and the countries is that this particular country's problem arises from a lack of permanent bodies of waters and rivers. Because of this, Saudi Arabia is forced to rely heavily on underground sources of water. A number of these underwater resources have been contaminated in various ways as a result of the Gulf War. This contamination was caused, in part, by agricultural-related seepage and Iraq's burning of oil wells during the conflict. Additionally, the Gulf War brought increased strain on water resources due to the bombing of water treatment facilities by U.S. troops. Further straining water resources, multiple oil spills in the Persian Gulf damaged Saudi Arabian desalination facilities. As can be seen, water serves as a catalyst for conflict among various Middle Eastern countries. As with neighbours in Texas, neighbours in the Middle East recognize the importance of protecting water rights in order to ensure continued access to this valuable resource. And, as might be imagined, conflict soon follows this recognition.

9

The Water over the Dam

The intent of the Reclamation Act of 1902 was to use public funds to irrigate the otherwise unfarmable deserts of the West. In 1903, six major irrigation projects were approved, and the Reclamation Service, which had been established by the act, began construction on the Truckee-Carson Dam in Nevada, to irrigate an area that was virtually uninhabited. In 1905, work began on the Roosevelt Dam, on the Salt River near Phoenix, Arizona. The dam was built with rock hewn from the canyon walls by a crew of Apaches not twenty years removed from Geronimo's band, and a few Mexicans and hoboes recruited from ranches and freight trains. When completed in 1911 at a total cost of $10.5 million, the Roosevelt Dam was the tallest rock masonry dam in the world. A solid wall bowed upstream against the river's current, rising 280 feet above bedrock. The Reclamation Service's ingenuity and resourcefulness and the graceful design of its dam astounded the world. The grasslands of the Salt River Valley, now assured an ample, inexpensive supply of water, were soon farmed, and Phoenix was on its way to becoming a desert metropolis.

After the Roosevelt Dam on the Salt River came the Arrowrock Dam on the Boise, which was even taller, the Elephant Butte Dam on the Rio Grande, and the Gunnison tunnel, dug through granite and shale to divert Colorado's Gunnison River into the Uncompahgre Valley. By 1919, twenty-six separate reclamation projects were in various stage of completion.

The newly irrigated lands were generally planted in alfalfa, wheat, cotton, or hay all low-value crops that were cheaper to raise in nonirrigated regions. Except for California lettuce and oranges, crops grown in the desert did not make the farmers rich. The overhead was high, and it was more difficult to make a living growing irrigated crops in the short-grass prairie than nonirrigated crops in the tallgrass prairie. In the good year of 1917, for example, a Department of Agriculture study showed that the Corn Belt farmers in Illinois cleared an average of $870; in Chester County, Pennsylvania, the farmers made $789; but on the irrigated land of Utah's Salt Lake Valley, farmers made only $417.

Figure: *Roosevelt Dam seen from downstream*

Nonetheless, the concept of reclaiming a desert was a potent lure, and in the first decades of the twentieth century, some of the best engineering graduates in the country gravitated toward the Bureau of Reclamation (as it was renamed in 1923). Projects were financed by a fund initially made up of revenues generated by the sale of public land in the West, and construction costs were meant to be paid back gradually, through the sale of water to farmers.

As the size of the projects increased, so did the cost of the engineering and construction. But the farmers made too little money to pay much for water, and it was soon clear that the cost of building a dam was unlikely to be recouped through water fees. The difference between the farmers' payments and the actual cost of the Bureau's forty year Columbia River project, for example, amounted to a 96.7 percent public subsidy by 1980. Dams were not built as moneymakers, though: they were built to bring farmers and civilization to unpopulated land. As a good landscape was one that lay plowed, planted, and stocked, so a good river was one that was dammed, channelled, and leveed. From an engineering standpoint, dams were wonderfully multipurpose construction projects, useful for hydroelectric power, flood control, and water supply as well as irrigation. Every canyon was a hole that was waiting to be filled, and water that ran to the sea without being used was considered wasted.

The dams generally had modest effects on the waterways until the 1930s. Floods were regulated and sediment was trapped, but the dammed rivers retained some of their natural character. Seasonal flow variation was less pronounced but nonetheless apparent; reservoirs were relatively small and affected the stream channel only for a limited distance. But in the 1930s the construction of large dams and the grouping of multipurpose projects within entire river basins became symbols of American engineering mastery over the landscape.

Figure: *The original Roosevelt Dam, completed in 1911*

The Tennessee Valley was the first fully integrated basin development in the country, and others soon followed. In 1936, the Bureau blocked the Colorado River with the Hoover Dam, which, like the Great Wall of China, is said to be visible from the moon with the naked eye. In Washington State, the Bureau concentrated on the Columbia River Basin. Eight dams had already been built on tributaries to the Columbia when, in the 1930s, the Bureau decided to dam the Columbia itself, on a glacial outwash known as the Grand Coulee. The Grand Coulee Dam was completed in 1942; ambitious projects were under way in the Central Valley of California; and before the century was half over, the American West had more and bigger dams on its waterways than any other region on the planet -and still the building continued.

Between government and private enterprise, roughly fifty thousand dams have been built in this country. Of these, a thousand or so are what engineers refer to as "major works" truly gigantic constructions holding back rivers that were once thought to be untamable: the Colorado, the Columbia, the Snake, the Tennessee. There are sixty dams on the Missouri River and its major tributaries, and twenty-five dams on the Tennessee. Why on earth did we build so many?

With the adoption of river-basin planning in the 1930s, a single project typically included dams, canals, and irrigation works from headwaters to river mouth, across 1,000 miles of terrain. During the Roosevelt and Truman administrations, several omnibus river bills authorised dozens of dams and irrigation projects at a single stroke. Economics mattered little, for while the irrigation projects might never pay off, the hydroelectric dams within the same river basin could generate the revenues to cover the losses.

Over a period of eighty-six years from 1905 to 1991 the Bureau of Reclamation and its predecessor built 339 reservoirs, 154 diversion dams, 7,670 miles of irrigation canals, 1,170 miles of pipelines, 270 miles of tunnels, 267 pumping plants, and 52 hydroelectric power plants. With the help of $18 billion in capital outlays, over 14,000 square miles of farmland received water, and the West bloomed and prospered. Water made much of this land intensely productive, and advocates of reclamation saw public irrigation as a miraculous font of free riches. There were, however, great costs.

Mammoth dams are able to exert total control over the downstream river. The waterways are manipulated for power, flood control, and water supply, and the result is a truly engineered river system, whose out-of-season high flows don't correlate with seasonal rains. Natural extremes of flow, water temperature, and sediment transport are eliminated. Instead, unnaturally rapid flow fluctuations or sudden periodic flow changes are often superimposed on a constant background discharge: the low flows are higher, the high flows are lower, and the cleansing and nutrient-filled annual flood pulse is eliminated.

When the Hoover Dam went up, fish that had evolved to survive in the widely fluctuating and muddy flows of the Colorado lost their competitive advantage. The muscled humps of the humpback and bonytail chubs, designed to allow these fish to hold position or make progress against currents that swept most fish away, were suddenly superfluous. Eight clams on the lower Colorado and an aggressive dredging program to redirect the riverbed have contributed to the extinction or threatened status of eight species of fish native to the Colorado. As a kind of compensation, channel catfish and rainbow trout were introduced, and have prospered.

Water takes much longer to move through a dammed river system, and nearly all the sediment and organic matter carried along by the water is permanently stored in the reservoirs. There is less organic matter in the water to fuel the aquatic ecosystem; worse, since natural river channels are maintained by a dynamic equilibrium of sediment deposition and erosion, water without sediment can trigger major riverbed erosion. When dams retain silt, erosion is accelerated for dozens of miles downstream. Below the Hoover Dam on the lower Colorado, for example, millions of cubic yards of sediment were scoured from the channel for 100 miles, and the channel slope was noticeably reduced. It took about twenty years after construction for the channel bed to be fully armoured and stabilised -that is, transformed into a boilerplate of bedrock and boulders. The water that flows in a stream is well mixed and well oxygenated. When water rests in a reservoir, it starts to stratify. Since cold water sinks and warm water rises, reservoirs generally stratify all summer long. Phytoplankton proliferate near the surface of the reservoir, releasing oxygen and keeping the water saturated with it. Little mixing occurs, and the sunlight does not penetrate below the top few yards. Dead phytoplankton settle to the bottom and are consumed by bacteria, which use oxygen to process the corpses, so water at the bottom of the reservoir is low in oxygen, high in organic matter, and high in salts as well. When autumn comes, the reservoir overturns, as the water on top chills and sinks and the water on the bottom becomes comparatively warmer and moves to the surface. The high-nutrient, high-salinity, low-oxygen water is now on top, while the low-nutrient, low-salinity, highly oxygenated water is on the bottom of the reservoir.

A dam is designed to release water either from the top or the bottom of its reservoir. Dams with deep-outlet discharges create streams with high nutrient levels, high salinity, and low oxygen during summer months, while in the fall these same streams are fed by nutrient-poor, low-salinity water with plenty of oxygen. During the summer, surface-outlet discharges create streams low in nutrients and salinity, and in the fall they create streams high in salinity and nutrients. Below a reservoir, thermal and chemical gradients in the river will persist for dozens of miles downstream. Silt builds up in whatever gravel beds remain, and the native fish are edged out by introduced species. Water quality shifts in odd ways, and the daily temperature variation in the stream disappears. Water temperature in a stream varies daily, by perhaps 7°F in the winter and 20°F in the summer. Below a reservoir, the daily temperature variation, if there is one, is generally less than 1°F.

These changes seem subtle to those of us who live on land, but to aquatic species they are dead obvious. Water temperature, for example, triggers egg hatching in many aquatic insects. Several species of mayflies require three specific temperature regimes to hatch: near-freezing water triggers egg development; a rapid rise in the water temperature precedes hatching; and minimum temperatures of about 65°F over several months stimulate nymph maturation. The constant temperatures below a deep-release dam may not provide the thermal cues necessary for an insect species to complete its life cycle, leaving few larvae or insects for the fish to eat.

When a watershed is dammed and lakes and reservoir-fed rivers replace the swift, turbulent, highly oxygenated water of a free river, the riverine species will begin to fade away. In time, the entire aquatic community of a dammed river will begin to resemble that of a lake ecosystem. Moreover, when rivers are dammed, the fish that mature in the ocean and swim inland to spawn in the rivers begin to disappear.

The Pacific salmon are among the last of the great anadromous fish runs, but there once were many fish species on both coasts which journeyed inland to spawn. In his 1705 *History of Virginia*, Robert Beverly states that "herrings and shad come in such abundance in the brooks of Virginia that it is almost impossible to ride without touching them." In 1851, William Stack reported that his father had seen "shad so thick in the Merrimack River that you could not put your hand in without touching some of them, and yet there were more alewives than shad." And when the eels ran in the streams of Onondaga, wrote Jesuit diarist David Zeisberger in the 1780s, a man could spear a thousand in one night. Eels are catadromous, spawning in the sea and maturing in rivers that empty into the Atlantic Ocean. When it's time to breed, eels lay on migratory reserves of fat totalling up to a third of their body weight, and swim thousands of miles to the deep of the Sargasso Sea, which lies south of Bermuda. There they lay their eggs, and die soon after. The young females return to fresh water; the males stay in salt or brackish water all their life.

Striped bass 5 feet long and weighing as much as 100 pounds once entered the streams of the mid-Atlantic states in great abundance, along with the shad, the alewives, and the Atlantic salmon; and the sturgeon were once so regular in their inland expeditions that the full moon of August was called the sturgeon moon. The largest of all freshwater fish, some sturgeon live for two hundred years, measure 20 feet long, and weigh close to a ton. (The sturgeon of Russia's Lake Baikal are said to reach 2,200 pounds and an age of three hundred years estimated from annual growth rings on their eardrums although it seems highly unlikely to me that a fish could survive from the reign of Peter the Great to the present.) Some species of anadromous fish die after a single spawning, but the sturgeon returns to the sea and makes forays inland to spawn every few years. Sturgeon eggs are nice as caviar, and their swim bladders were once used to make isinglass windows (that roll right down, in case there's a change in the weather).

The sturgeons of the Great Lakes were so large that the spawning runs were easily obliterated by commercial fishermen at the turn of the century. As early as the late 1700s, overfishing began wiping out the Atlantic salmon as well. The survivors are much smaller today than they were two hundred years ago. An early writer reportd that 40-pound Atlantic salmon were occasionally taken on the Grand Codroy River in Canada. In this generation, the average salmon caught there tips the scale at 4 pounds, with the largest fish, year after year, weighing in at less than 25 pounds. But the anadromous species that commercial fishing didn't take care of, dams did. The rivers of the East powered the early Industrial Revolution, with dams built by entre preneurs running mills of all sorts. Dams proliferated on the smaller streams, and in the first half of the 1800s industrialists commandeered the larger streams, while lumbermen dammed the rivers to create high flows for enormous log drives. The New England states passed laws to protect the salmon to no avail, since the fish could not swim past the dams. By the 1870s, the Atlantic salmon had virtually disappeared, while there were 433 laws on the Maine books to control and preserve the fisheries.

In 1890, perhaps haunted by the recent disappearance of the Atlantic salmon, Washington State's first legislature passed a law requiring that fish-passage devices, such as fish ladders, be built on dams "wherever food fish are wont to ascend." Federal fisheries laws were also passed in the last two decades of the century, and these also required that dams permit some way for migrating fish to pass. Nonetheless, by 1940 eight unladdered dams had been built on the Yakima River in Washington's Columbia Basin, and they had decreased the annual salmon run from six million to nine thousand fish. The Grand Coulee Dam effectively closed more than 1,000 miles of spawning streams in the upper Columbia Basin, destroying the legendary "June hog" King salmon run in that region.

Salmon, the most studied of the anadromous fish, are beautiful swimmers, who glide effortlessly on unseen currents and can accelerate faster than a car for short distances. The Chinook, or King salmon, is the largest of the Pacific salmon, and will occasionally weigh in at 125 pounds, with a 5-foot-long body; the other four species range down to 20 pounds packed into a 2-foot torpedo. And they used to be numerous. The Columbia River once had annual runs of an estimated sixteen million salmon.

Young salmon live in the ocean, feeding on zooplankton, small fishes, and squid. They follow the prevailing North Pacific currents in sweeps of 1,000 miles, travelling regularly each year in migratory patterns that depend upon the currents, the climate, the contours of the ocean bed, and possibly even the sun and stars. Salmon have good eyesight and can also detect pressure waves, using currents to navigate; they may even navigate using the earth's magnetic field. Fish migration is now believed to be based on exploration and learning, and as new habitats arise, the fish rapidly exploit them.

When a salmon is about four or five years old and has travelled 10,000 miles or more in its circuits of the great oceanic grazing grounds, it puts on weight and heads inland to spawn. Some salmon spawn in the spring, others in the fall; some swim as far as 2,000 miles inland to reach their natal stream. Salmon, like many fish, follow their noses. They distinguish between species and sex of other salmon by scent, and use olfactory landmarks to determine their location. They remember the smell of the vegetation and aquatic residents of the stream they grew up in, and smell their way back to breeding territory.

Although the salmon's life path was never easy, it has become unspeakably difficult to be anadromous today. If a salmon needs to ascend (for example) the Columbia to find its natal stream, it has to start with the Bonneville Dam. Fish mill about at the bottom of the dam waiting their turn at the fish ladders, bottlenecked in the odd-quality water straight from the reservoir. At the top of the Bonneville ladder, they are confronted with 40 miles of warm, placid water, very different from the cool, oxygen-filled waters of the rapids and river currents it replaced. After the Bonneville Dam comes the Dalles and then the John Day and the McNary each a major barrier to fish migration and each a major lake to swim through. Next come the Priest Rapids Dam, the Beverly Dam, and the Rock Island Dam, for the Columbia is no longer a river but a series of slackwater lakes.

Every dam that went up inundated a stretch of spawning stream on one side and eroded the gravel on the other. When too many dams are built on a river and too many downstream miles of riverbed are scoured, there are too few miles of gravel beds left for riverine fish to make nests in. The salmon, having made the journey, will dig their redds one on top of the other if necessary, each fish laying and fertilizing perhaps five thousand eggs.

Salmon hatch after two months of incubation in their gravel nests. The hatchlings live in the gravel until the yolk sacs attached to their bellies are consumed, after which they are recognisable as little fish. The fingerlings hide behind logs and under ledges, grab their food quickly, and retreat for cover. They swim near the bottom of the stream, where the current is weakest, travelling alone and conserving energy by resting in eddies. They come to know the stream's hiding places and its predators, and when the water runs low the young salmon will dive into the gravel rather than swimming unprotected into the current. They are surprisingly skillful at making their way through the gravel substrata, and have been known to follow underground watercourses into wells and springs, making it seem as though salmon fell from the sky. The fingerlings live in their river for as long as a year before becoming smolts, a silvery transformation that allows them to survive in salt water and immediately precedes their swim to the sea. Which is when things get extremely dicey.

A highly stratified reservoir can have surface temperatures that are lethal to juvenile salmon, while the cooler subsurface water is usually low in oxygen. Reservoirs are fine homes for lake fish, but they can become impassable barriers for juvenile salmon. Once the smolts get through the reservoir, they must pass through the hydroelectric turbines, which can be a bruising experience. In low-flow years, they may have to go through the turbines of seven or eight dams before they reach the ocean, and as many as 95 percent of the smolts will be macerated en route.

In high-flow years, the juveniles may be able to avoid the turbines, but the water that goes over the dam's spillway captures air and plunges it to substantial depths, mixing too much nitrogen into the water. A young salmon swimming in water supersaturated with nitrogen can die of the fish equivalent of a massive attack of the bends. Nitrogen bubbles form under their skin, their eyes bleed, sometimes their internal organs explode. In 1970, when the nitrogen levels in the water of the Columbia and Snake Rivers reached 143 percent of air saturation, the National Marine Fisheries Service estimated that 70 percent of downstream migrant salmon and steelhead trout were killed by nitrogen supersaturation before they reached the ocean.

The Pacific salmon's habitat has also been degraded by deforestation. Streams in deforested land are warmer, and the quiet pattern of pools and riffles created by fallen logs disappears, reducing food sources for the fingerlings. The salmon coevolved with the beaver, and until John Jacob Astor made his fortune in beaver skins from the Columbia River Basin, beaver ponds were part of the salmon's early life and part of its success as well. It is not surprising fish ladders and legislation aside that the salmon population started to fall off rapidly as the Pacific Northwest was dammed.

The solution to this complex environmental problem was technical: build hatcheries. The first salmon hatchery in the United States was built in 1871 on the Columbia, and between 1900 and 1930 the number of fry annually released by the state increased from twenty-five million to ninety million. So great was the government's faith in hatcheries that when a dam was built on the Olympic Peninsula's Elwha River in 1915, the Elwha hatchery was opened in lieu of emplacement of fish ladders, and in the next few years seven more hatcheries were built to compensate for the fish runs lost to dam construction on the White Salmon, Chehalis, and Elwha Rivers. Initially, two million eggs were collected annually to hatch in tanks, but within a few years the pool below the Elwha dam was empty of fish. The Washington Department of Fisheries abandoned the Elwha hatchery in 1922.

In the 1930s, Canadian fisheries biologists showed that substantial releases of hatchery sockeye increased neither the commercial catch nor the number of fish spawning in the wild. Canada closed all of its Pacific salmon hatcheries, and a number of American hatcheries on the West Coast were closed. But the Washington Department of Fisheries expanded its network. In 1958, the first of twenty-five Washington fish farms was built. By 1966, the program, which used natural lakes to raise large numbers of confined fish, was abandoned, because almost no farm fish were coming back to spawn. But to prepare the selected areas for the farmed salmon, all the native fish had been poisoned, so the failed fish farms, in addition to costing millions of dollars, removed dozens of wild salmon runs.

Hatcheries continue to release many millions of salmon every year, and the salmon returns have continued to decline. In 1992, a study conducted by

Terry Divietti, a Central Washington University psychology professor, provided an explanation. When fish are raised in a hatchery, they learn far less than the tiny fingerlings that grow up in the riverbed. Hatchery fish in concrete tanks have no threats in their life; they Rock to the surface for food, and swim in packs the rest of the time. By contrast, fish that grow up in a stream learn to survive dozens of predators, and will hide, dart out, and zip back under cover to eat. They swim alone. They know how to use rocks and logs to conserve energy, having honed their swimming technique in the variable, complex flows of a stream. Hatchery fish learn none of these lessons in their tank. Releasing hatchery fish to the wild may be the piscine equivalent of sending a well-fed adolescent who has watched a lot of television into the woods to survive on his wits.

The other problem with hatchery fish is genetic. In the last few decades, it has been found that stream fish are much more genetically dynamic than anyone had ever imagined. There once were roughly 1,000 breeding stocks of salmon, of which 106 are now extinct and 314 are threatened or endangered. A human being offers up a few genetic carriers to the world; a salmon offers up thousands of genetically unique offspring, and the few that make it through to adulthood are likely among the very best of the lot.

Therefore, salmon species adapt relatively quickly to local conditions. Genetically speaking, each stretch of river is home to its own strain, and every adult that returns is the pick of a very large litter. Hatcheries provide the protection that salmon are designed to do without, so hatchery-raised fish are genetically unculled. Releasing millions of hatchery fish plays havoc with the wild salmon, for both compete for the same food sources.

Hatchery fish are much less likely than wild salmon to survive to adulthood, but they do apply pressure on the surviving wild stocks. Thanks to dams, habitat degradation, overfishing, and hatcheries, the Pacific salmon has now disappeared from over 40 percent of its former range in Washington, Oregon, Idaho, and Montana. In 1994, despite a billion-dollar salmon recovery program, only about two and a half million salmon returned to spawn in the Columbia and its tributaries (two million hatchery fish, half a million wild), and the numbers continue to fall.

When anadromous fish swim up rivers to spawn, they provide an input of nutrients gathered at sea which are enjoyed by the entire food chain. Anadromous fish are consumed by carnivores, like wolverines and eagles, and omnivores, like bears and people; and the plants benefit as well, because fish concentrate phosphorus, one of the critical nutrients for plant growth, in their bones. The fish-laden feces of larger animals and the offal and bones left on land once delivered a nutrient element that present-day inland soils are often short of. When the spawning runs of fish declined, there was less flesh to sustain the meat eaters, and the green plants at the bottom of the food chain had less phosphorus to grow on, so the deer and elk had fewer new shoots to nibble on. No catastrophe, mind you, but it's harder to be fat and happy when the fish don't run.

Mussels, Gators, and the Corps

About one-third of the world's freshwater mussel species some three hundred of them are found in the waterways of the United States, and their names are a colloquial smorgasbord: purple pimplebacks, three-horn wartybacks, fat pocketbooks and spectaclecases, rough fatmuckets, pink heelsplitters, turgid blossoms, and Waccamaw spikes. Measuring up to 10 inches long, freshwater mussels burrow into the river bottom. Their shells lie slightly ajar when they feed, and their mantles frills of golden flesh reach up into the water to filter out the impurities for food, cleaning the flow. Most species of mussels are slow-growing and long-lived: many take a half dozen years to reach maturity and live for periods varying from fifteen to fifty years. Some genera prefer swiftly flowing streams with gravel bottoms; other are found in muddy lake beds or sluggish, silty rivers. In the United States in particular, freshwater mussels exploited their niche with great success, and dense beds of mussels once covered the bottoms of streams, lakes, and rivers, clarifying the water.

Mussels are not mobile as adults, but when they are young they live for a while on the fins or gills of fish. Larval mussels, called glochidia, are discharged from the female mussel's marsupiallike pouch into open water. The larvae need to attach themselves to a suitable fish quickly, so mussels have evolved a number of strategies to lure fish into serving as obligate hosts. Some wave attractive body parts to draw fish near, while in other species the larvae themselves are easily mistaken for worms or minnows. Once safely encysted on a fish, the glochidia grow and eventually undergo metamorphosis into adults; that done, they drop off the fish and begin an independent existence in burrows at the bottom of the waterways.

This arrangement allowed the mussels to use the motive power of fish to spread throughout entire watersheds, leaving no likely habitat uncolonised. Inevitably, certain mussels evolved a dependence on specific host fish. The yellow sand shell, for example one of the largest and most beautiful of the freshwater mussels in the Mississippi drainage system, renowned for producing large, lustrous yellow-tinged pearls confines its attention to the gar, a once-plentiful species that became scarce when commercial fishermen targeted it as a destroyer of commercially valuable fish. (And not without reason; it is estimated that a single gar will eat 10 pounds of flesh for every one of its own, and then some.) As its numbers dwindled, so did the population of yellow sand shells.

Some mussels adapted to depend on specific fish, and some fish adapted to exploit the mussels. The sheepshead is a huge stout-jawed fish with round plates in its jaws and throat, perfect for crushing mussels and separating the flesh from the shell. When mussels are shattered this way, thousands of glochidia are released, and they encyst themselves on the sheepshead that has just eaten their parent.

Freshwater mussels, like their cousins the saltwater oysters, make pearls. The various species of mussels in American waterways, with different shades of nacre lining their shells, produce pearls in a wide range of colours, from white to faint shades of ivory, pink, yellow, and salmon, on through purple, copper, red, bronze, and steely blue. The freshwater pearls most familiar today are the grayish-white Japanese pearls cultured in the mussels of Lake Biwa, in central Honshu. Lumpish and baroque, they barely suggest the unsurpassing loveliness of the finest freshwater pearls. Round and with luster that makes them seem almost translucent, the best of the freshwater mussel pearls are said to be more luminous than the finest pearls from saltwater oysters. The North American Indians gathered freshwater mussels for food, pearls, and wampum, leaving small hills of shells on the shores near prolific mussel beds. With the removal of the Indians, the freshwater mussels multiplied undisturbed, and the waterways teemed with them. The mussels and their pearls had a long period of uninterrupted growth.

Although mussels are not physically active, when their shell is ajar they will clamp down so hard on anything that touches their mantle that they can be pulled right up out of the riverbed on the end of a stick. In 1857, a carpenter who was randomly killing mussels found an exquisite pink pearl of 93 grains in a mussel pulled out of Notch Brook, near Paterson, New Jersey. (A standard measure for pearls, a grain equals 64.799 milligrams. There are 3.086 grains in one carat, which is 200 milligrams.) Nearly perfectly round and over a half inch across, the carpenter's Brook find was soon named the Queen Pearl and bought for $1,500 by Tiffany & Company, which then sold it at a two-thirds markup to Empress Eugénie of France. Fortune hunters immediately descended on Notch Brook and the surrounding waterways. The furor abated in a few seasons, after about $50,000 in pearls had been found and nearly every mussel within miles had been opened, but pearl fever broke out about every ten years after that, for the remainder of the century.

According to George Frederick Kunz *Gems and Precious Stones of North America* (1892), farmers often hunted for pearls in their spare time. In the late 1800s, a single pearl from Montpelier, Vermont, brought $300 (the price of nearly 15 ounces of gold), and most of the streams in the state are thought to have lost their mussels in the excitement that followed. A pink pearl from Murfreesboro, Tennessee, sold for $150, single pearls from Texas streams sold for up to $250, and ten thousand people were pearling in Arkansas in the last decades of the 1800s. In the 1870s, two anthropology professors excavating an Indian mound in Ohio's Little Miami Valley found a cache of about sixty thousand freshwater pearls, drilled and undrilled -many of them spherical, and varying in size from 1/10 to 1/2 inch in diameter. The pearls were decayed and of no commercial value, but in the wake of their discovery many Ohio streams were carefully raked and scraped. In 1878, the mussels of Waynesville, Ohio, gave up about $3,000 in pearls.

In the summer of 1889, a number of dark pink, purplish red, copper red, and metallic green pearls were found in the streams of southwest Wisconsin, and three months later $10,000 worth of Wisconsin pearls had been sold in New York City. A set of ninety-three Wisconsin pearls fetched £11,700 in London in 1890, while a necklace of thirty-eight freshwater pearls from the United States sold in 1904 for 500,000 francs in Paris. After eight years, Wisconsin streambeds had yielded about $300,000 in pearls, and the mussels in many of Wisconsin's waterways were nearly exterminated.

The Mississippi was once home to many of the finest pearl- producing mussels as well as nearly all the species of mussels whose shells could be used for pearl buttons. It was the pearlbutton business that nearly did in the Mississippi mussel beds. In 1891, a German named Johannes Boepple pioneered the shell-button industry with a small pearl-button factory on the banks of the Mississippi in Muscatine, Iowa. Within three years, mussel collectors had harvested nearly 10,000 tons of shells from a section of river near Muscatine about 1 1/2 miles long and 300 yards wide, and when that bed was depleted the gatherers extended their quest. By 1897, 167 miles of the river were being harvested.

With ample raw materials and a ready market, pearl-button factories sprang up throughout the region. From the start, these factories were supported by itinerant shellers, who followed the mussel frontier in shantyboats. Mussel gatherers lived and worked in makeshift riverside camps, and temporary cutting plants were set up until the beds gave out. Although the shell factories accepted only certain species of shells, pearls provided roughly a third of the sheller's income, so the shellers would steam open their entire catch regardless of species, in large, flat pans. After removing the pearls, they would discard the meat and wash, sort, and market the shells. When a bed was first harvested, mature mussels with large shells were taken. After a few years, the average size decreased, and eventually only very small shells were left. Gatherers had to collect two or three shells to get the value of one, and as the supply fell the price rose, stimulating the shellers. Finally, when an area had been thoroughly depleted and the shellers moved on, the local population continued to work the depleted mussel beds periodically for extra cash, so the beds were unable to re-establish themselves. At the turn of the century, there were sixty button factories along the upper Mississippi in Iowa, Illinois, Missouri, and Wisconsin, employing some two thousand workers. Most of these shops were "saw works," where rough blanks were cut and then shipped to New York and other Eastern cities for processing into completed buttons. In the winter of 1898-99, a shell buyer in Le Claire, Iowa, shipped 1,000 tons of shells to New York, representing a harvest of perhaps ten minion mussels. The adoption of automatic machinery in 1901 accelerated the pace. The annual output of completed buttons shot up from 1.6 billion in 1904 to an all-time high of 5.7 billion buttons in 1916, by which time shelling had spread throughout a large part of the Mississippi Basin and into a few streams that drained into the Great Lakes and the Gulf of Mexico.

The mussels were gathered commercially by a tool called the crowfoot, made of an iron bar with up to a dozen four-pronged hooks attached to it. The crowfoot was dragged along the riverbed by a boat, and the mussels would snap their shells shut on the hooks. Dragging a bar bristling with iron hooks repeatedly along the riverbed is not good for growing mussels, and the shellers were persistent. In 1907, the U.S. Bureau of Fisheries guaranteed the mussel supply and funded a program of artificial propagation. The commercial propagation of mussels began in 1912, concentrating on the upper Mississippi, the Wabash in Indiana, the Ohio River, and the White and Black Rivers in Arkansas. Mussel glochidia were put into tanks with host fish, which were released to the rivers with the mussels safely encysted. This program dovetailed nicely with another bureau project the fish rescue program, begun in the early 1900s. Each spring, the Mississippi would overflow its banks to spread into adjacent forests. River fish would spawn in these lowlands, and when the water receded millions of fingerlings were stranded in the slowly evaporating backwaters. Believing that fingerlings returned to the river would reach maturity and be caught by commercial fishermen, thus enriching the markets of the nation, the bureau hired crews to bring these landlocked fish back to the Mississippi. Before the rescued fish were released, many became encysted with glochidia, so the fish rescue and mussel propagation programs reduced overhead costs for each other.

With the mussel propagation program, the fisheries bureau hoped to save an economically important, highly competitive industry from self-destruction, but they did little to stop the use of the pernicious crowfoot; not until 1933, by which time the industry was moribund, did they state that if its use was allowed to continue the mussel beds would not recover. Between 1914 and 1920, the Mississippi produced 35,000 tons of shells a year and supplied about two-thirds of the world's buttons. At the height of the boom, the button business employed about twenty thousand people in the fisheries and factories, manufacturing $12.5 million of buttons a year. But it was the beginning of the end. In 1930, the Bureau of Fisheries abandoned its mussel-stocking program, because it was an utter failure, and urged the five states along the upper Mississippi to repeal their mussel-protection laws the dying industry, instead of being hobbled, should be given free rein. "It appears that the mussel fishery of the Mississippi River is doomed to economic exhaustion," wrote the commissioner of fisheries in 1932, and the bureau's official stance was that all the remaining mussels should be harvested.

Harvesting had caused local disruptions in the mussel beds, but when the waterways were engineered for transportation and flood control, the fish populations fared as poorly as the mussels. At one time, the Mississippi River supported abundant populations of the continent's largest freshwater fish: the white sturgeon (*Acipenser transmontanus*), a living relic covered with bony plates, topped out at 1,800 pounds; the paddlefish, a longsnouted, weird-looking fish, the sole survivor in this hemisphere of an ancient family (there is only one living close relative, in China) routinely weighed 100 to 200 pounds. Sturgeon and paddlefish were bottom feeders, while the gar pikes and bowfin, two oversized relics of ancient, purely American families, were voracious predators. Catfish, found everywhere but in swift mountain streams, grew to great size on the Mississippi riverbed: mud cats, channel cats, blue cats, and Mississippi cats often weighing over 100 pounds were caught, as were buffalo fish, blue suckers, and the sheepshead that fed on the mussels.

The lower Mississippi also supported dense populations of alligators, until the mid-nineteenth century. Another keystone species, alligators increase the complexity of swamps and marshes by building large mounds as nests and digging gator holes to overwinter in. In 1826, John James Audubon wrote that on the banks of the Red River as it neared the Mississippi, alligators "could be seen along the shores by the hundreds, or on immense rafts of floating or stranded timber, the smaller on the backs of the larger, all groaning and bellowing like thousands of irritated bulls about to fight." If an alligator hollowed out the same hole for several years in a row, the hollow would become a small pond. During droughts, gator holes sometimes provided the only open-water habitat available to amphibians, fish, and the predators that depended on them.

Alligators also maintain open-water habitats by tearing out the aquatic and emergent vegetation. In the spring, a female alligator will scoop up mud in her jaws, mix it with mouthfuls of vegetation, and build a mound 5 to 7 feet across at the base and roughly 3 feet high, in a shaded area near the water's edge. She lays between twenty and seventy eggs in the hollow on top of the mound, covers up the eggs, and tamps down the top of the mound with her feet to smooth it out. She then guards the mound faithfully, until late summer or early fall, when she's alerted by the peeping sounds of baby alligators and carefully digs up the eggs to free her children, 8-inch-long miniature alligators emerging from 3-inch-long eggs. Carrying them gently in her mouth, she takes the hatchlings to the water and watches over them until they are able to manage the swamp by themselves.

Thousands of alligator mounds once served as dry spots in the swamps along the Mississippi, providing nesting areas for turtles and colonisation areas for plants, which the birds could nest in. Trees and shrubs root on old alligator mounds, so the Mississippi alligators often initiated the transformation of a marsh into a lowland forest. Between flooding and alligators, the entire length of the Mississippi River was once fringed by a complex edge of swamps and backwaters. Pockets of sedges, pond weeds, and millets fed huge assemblages of waterfowl. From the wild rice marshes of Minnesota to the coastal wetlands of the Delta, the Mississippi flyway, fed by the river's riparian edge, was an aerial highway. It was used by tens of millions of migratory ducks, which followed the seasons between summer nesting grounds in northern Canada and wintering areas near the Delta, while many more millions of birds flew it en route to Central and South America.

Floods are an annual occurrence on the Mississippi, and the broad hardwood forests of the lower Mississippi Valley were inundated seasonally. Floods expanded the habitat for many aquatic species, creating prime acreage for foraging, spawning, and a safe nursery for those vulnerable first weeks after hatching. Many fish made regular migrations to use flooded forests to spawn, and annual floods provided an enormous nutrient pulse to the land.

The first levees on the Mississippi were built in 1699 by individual landowners to confine the floodwaters, but work on the river was generally restricted to removing snags, bars, and caving banks, until the waterways were used more intensively for transportation. The first steamboat was built in Pittsburgh in 1811; eight years later, there were 191 steamboats plying the river; by 1833, more than 1,200 cargoes were unloaded at New Orleans. The cost of shipping freight between Cincinnati and New Orleans fell from 9¢ a pound in 1814 to less than half a cent in 1828, bringing an era of unprecedented prosperity to the river. Steamboats contributed to the widespread clearing of the riverside forests, for most of these vessels were wood-fired until the 1880s, when coal took over. The riparian zone was always the first to be cut. This cutting, combined with agricultural clearing, contributed to bank instability, while increased sediment loads and runoff from agricultural land increased bank caving as well.

In the early 1800s, much of the Midwest, particularly areas of Iowa and central Illinois, was swampy and poorly drained. This swampland was infested with malarial swarms of mosquitoes, but when ditches and channels were dug to shunt off the water, the marshes became rich bottom land fields. The work was done with horse-drawn slip scrapers, gangs of men with shovels, or enormous ditching plows, the largest of which required sixtyeight oxen driven by eight bullpunchers. Later, the dredging was done by huge floating machines, which were built on site, and most of the channels required regular dredging. The dredging of waterways containing mussel beds killed mussels directly, while mussel beds downstream were clogged with silt.

Channelling has drained more than 200,000 square miles of wetlands across the country, acre by acre and farmer by farmer. The states bordering the upper Mississippi, the Ohio, and the Missouri Rivers have lost over 80 percent of their wetlands. What the beavers built and left behind, the farmers ditched and drained. The land was channelled, and the rivers were channelised as well not just to drain land for agriculture but also to improve water transportation and to speed the water away to reduce local flooding (which increased the flood pulse downstream). Swamp reclamation and navigational improvements caused rivers to flow more rapidly to the sea, river levels fluctuated more than before, and both of these changes were unfavourable to most fish populations. Moreover, artificial drainage increased the efficiency with which the available moisture was collected and carried to the waterways, so deepening a river channel or constructing new drainage works eventually lowered the local water table.

Much of the channelisation was carried out by the Civil Works Branch of the Army Corps of Engineers. The Corps was first formed in 1802 and undertook the improvement of inland navigation in 1824. As the country grew, so grew the Corps. Soon it was not only digging ditches and canals for drainage and transportation but building dams to manage floods, irrigate land, and generate hydroelectric power to sell. But its primary goal was to turn every river into a highway for barges, and it has done an excellent job. The Corps currently maintains more than 19,000 miles of improved inland and intracoastal waterways, and also local flood-control projects incorporating over 350 dams (which sold nearly 70 billion kwh of electricity in 1992), along with more than 9,000 miles of levees and floodwalls and 7,500 miles of improved channels; while the Soil Conservation Service has channelised an additional 8,000 miles of rivers. The larger projects are part of river basin planning, but the smaller projects have often been initiated by people that hope to profit from them. If, for example, a rancher grazed cattle on a floodplain that was underwater half the year, he might ask the local congressman to arrange a feasibility study; if the engineering surveys and cost/benefit ratio looked practical, contracts were awarded and the dirt began to fly.

Even as late as the 1960s, the natural water cycle was so far removed from the minds of the Army Corps of Engineers that it considered the riparian zone the ecotone between water and land an unacceptable waste of water. The plants that live along a stream and suck water from the saturated zone are collectively termed phreatophytes (from the Greek for groundwatered plants). In a particularly wretched episode of engineering run amok, the Corps started stripping the phreatophytes from the riverside in arid regions to conserve the streamflow. Along the Rio Grande River in Texas, 95 percent of the whitewing dove habitat was removed when the riparian zone was rased, and four Million doves disappeared. Along the Gila River in Arizona, the lush riverside habitat supported doves, quail, pheasant, peccaries, deer, and even a few bears, while the trees cooled the water and supported generous populations of smallmouth bass and other fish. By the time the Corps bulldozers were through, the river had been transformed into an overheated channel that supported neither fish nor game.

The purpose of channelling wetlands and rivers was twofold: to dry up the land and to improve water transportation, and on both counts the Corps was remarkably successful. Wetlands became farmland, the river bottoms were deepened, and the river biota changed as well. Channelisation and dredging removed the pools and riffles from the streambed, simplifying the habitat. As a rule, when this happens the larger fish disappear and tiny minnows, shiners, and darters remain. In a 1960s study of the Tippah River, the Mississippi Game and Fish Commission found before channelisation a total standing crop of 877 fish per acre, weighing a total of 241 pounds. After channelisation, 99 percent of the fish were darters, shiners, and minnows. The total standing crop had nearly doubled to 1,498 fish per acre but the total weight per acre was only 5 pounds. In studies across the country, stream productivity was seen to crash when the streams were channelled.

In 1879, Congress established the Mississippi River Commission to improve the river for navigation. To cope with the instability of the river's banks, a large-scale system of levees was initiated in 1884 that restricted the normal migration of the channel and prevented the creation of oxbow lakes. To improve navigation, the Corps set out to construct a channel 9 feet deep and 250 feet wide at low water between Cairo, Illinois, and the Gulf of Mexico. Dredging began in 1896 and sent a steady flow of sediment downstream, disrupting the riverbottom community. In 1913, the first dam was built on the upper Mississippi, at Keokuk, Iowa. Extolled by a writer in *Iowa Magazine* as "the most colossal engineering feat ever attempted, not only rivalling, but actually surpassing, the ancient pyramids and the sphinx on the Nile," the 53-foot-high Keokuk Dam backed up the river for nearly 65 miles. By the end of the 1930s, twentysix locks and dams had transformed the upper Mississippi into a series of lakes.

After the Keokuk Dam was built, the U.S. Bureau of Fisheries kept a close watch on the riverine life. The commercial fish harvest from the Keokuk impoundment peaked at 900 tons in 1917, and began a long downward slide that reached bottom in the 1930s, when fishermen reported their take as 1:20 tons. By then, the mussels were all but gone, too. In Lake Pepin, along the Minnesota-Wisconsin boundary one of the bureau's centres of artificial propagation the mussel take fell from 3,000 tons of shells in 1914 to 200 tons in 1919, when sections of the lake were closed to allow the mussel beds to recover.

Meanwhile, domestic sewage, industrial waste, and eroded silt were entering the upper river in ever increasing amounts. Minneapolis and St. Paul added sewage, while deforestation, plowing, and the draining of wetlands increased the silt load in the water. Sewage combines with fine silt and is carried farther downstream than it is in clear water, so local waste disposal suddenly had a regional impact. When a dam is built, the sewageladen silt that would have washed downstream settles to the bottom of the reservoir, where it breaks down slowly and severely depletes the oxygen content of the water.

In 1927, the floodwaters of the Mississippi topped the embankments, and more than 26,000 square miles of land in Mississippi, Louisiana, Arkansas, Missouri, Tennessee, and Kentucky were inundated. Entire towns were underwater; roads, rail, and telephone lines were out of commission; and 214 people were drowned. As a result of this flood, engineers took a closer look at the river. The Flood Control Act of 1936 brought about the modification of an estimated 16,000 miles of the Mississippi's main channel and its tributaries. The main channel was widened from 250 to 300 feet, and further dredging deepened it to 12 feet an ongoing effort that has ultimately extended from the Gulf to Minneapolis-St. Paul. Between 1932 and 1955, the lower river was shortened a total of 152 miles by means of cutoffs dug through its meanders, and a further 54 mile shortening was planned. These cutoffs were constructed to reduce erosion and make the river's energy available to transport sediment, and they had a dramatic effect on riverine life. To keep the river in its new banks, the new channels had to be continually dredged for more than ten years, and between 1932 and 1955 over 1,300 million cubic meters of sediment were sent down the river. Finally, dozens of dikes and miles of revetments were constructed between 1945 and 1970, to help nip meanders in the bud.

A natural river writhes like a snake, throwing off oxbow lakes as meander loops slip downstream. But the flow in the lower Mississippi River has been transformed from a broad band of shifting riverbeds to a single, narrow, deep, swift channel. Walled in by 1,870 miles of main-stem levees, the Mississippi has been isolated from the surrounding floodplain, the riparian zone replaced by mattresses of concrete blocks. Fish that once relied on their annual visit to flooded hardwood forests for food, spawning, and nursery areas now have fewer such sanctuaries. The levees prevent the water from spreading out and dropping its silt, so the organic matter that once replenished the terrestrial ecosystem is now flushed out to the ocean, where it settles onto the continental shelf. The absence of silt results in coastal recession and subsidence and a shift in nutrient cycling: the nutrient elements that once fuelled the estuaries are now deposited in deep water. The Mississippi, dammed, channelled, and leveed, doesn't move much anymore. The rafts of logs and the alligators are confined to remote backwaters, the great fish that the river was once renowned for have slipped into legend, and the indigenous mussels, which once filtered the impurities from the water, are just about gone.

The upper Mississippi and the Missouri were once characterised by high turbidity, wide seasonal fluctuation in flow, and a wide, braided channel that constantly changed location. Bank erosion was continuous, and the channels shifted every flood season. Today there are twenty-six dams on the Mississippi and sixty on the Missouri. The Mississippi is a highway for barges, constricted to a dredged, leveed channel. Freshwater mussels are now the most widely endangered family of organisms in the United States, and malacologists (those who study mussels) say that the mussels are poised on the brink of a major extinction. One in every ten mussel species is thought to be extinct already, and two of every three of the remaining species are threatened. And when the great freshwater mussel beds disappeared, the water flowed unfiltered to the sea. One of the earliest problems of human civilization was keeping feces out of the water supply. Almost without exception, cities were sited near lakes, streams, or springs, which supplied the citizenry with fresh water. As the population grew, runoff and daily dumping polluted the surface water, while waste from outhouses, cesspools, and stables seeped down to the groundwater. Before long, the lake or river would be undrinkable and the wells and springs tainted. Feces and water are both vectors for the transmission of many diseases, and cities are dangerous places to live unless clean water can be piped in from remote locations and wastewater can be piped out in sewers.

The technical problems of water supply and sewer systems were solved early. By 2500 B.C., the Sumer and Indus Valleys had huge irrigation works, and the cities that grew from those well-watered fields were fully plumbed as well. Aqueducts brought fresh water from the mountains into the city to supply public fountains, baths, and latrines, while sewers shunted wastewater out of the city and back into the waterways. Some private Sumerian homes had water-flushed latrines that connected to the sewer system, a convenience that didn't become commonplace until four millennia had passed.

Roman aqueducts and sewers have long been admired as one of the marvels of the ancient world. In 97 A.D., the former governor of Roman Britain, Sextus Julius Frontinus, was appointed superintendent of Rome's aqueducts, and his *De Aquis Urbis Romae* is a comprehensive tour of the triumphs of Roman waterworks. Instead of drinking from the polluted Tiber or tainted local wells, the Roman population drank water brought to the city from the forest streams of the Apennines, in aqueducts built of local stone and lined with cement. By the time Frontinus wrote, the Romans had built 255 miles of aqueducts carried on arched bridges across valleys, tunnelled through rises, and supported on structures of solid masonry. Within Rome, water was distributed through lead pipes and hollowed-out logs to public baths and fountains, to 144 public latrines (some of which had 20 seats to a side), to shops, and to dozens of private homes. Water use per capita was about 300 gallons a day, roughly three times what it is now in the United States.

After the Roman Empire fell, Europe is said to have gone unwashed for a thousand years. With the ascendancy of Christianity, the wealth that the Romans had diverted into the construction of aqueducts and sewers was spent on cathedrals and their bureaucracy, for the Church was concerned with loftier matters than its flock's temporal existence. Ironically, monasteries were the only places where plumbing was regularly found in the Middle Ages: many had piped water systems and latrines connected to sewers. City wells were invariably fouled and the streets were even fouler, but people tithed, prayed, and bought their drinking water from carts which hauled it from springs outside the city limits and they washed only rarely, with water from tainted wells. Which became progressively more contaminated, for Christians preferred to be buried in the sanctified ground of the churchyard, creating layer upon layer of putrefying parishioners, all leaching down to the groundwater? Over time, urban churches laid hundreds of thousands of believers to rest in a small area. The Scottish Journalist Basil Hall writes in 1820 of churchyards ankle deep with "rank and offensive mould, mixed with broken bones and fragments of coffins," while the water of city wells near churches grew redolent with putrescine, cadaverine, and other pungent organic compounds associated with decomposition.

At the time of Independence, Boston, New York, and Philadelphia were drawing their water almost exclusively from wells, cisterns, and springs. As in the Old World, Americans buried their dead within city limits (the churchyard of New York's Trinity Church, for example, held a hundred and sixty thousand graves by 1830). The contents of cesspools and outhouses leached into the groundwater, and city wells went from bad to worse. By 1809 the water from New York City's public pumps was so foul that horses balked at drinking it; by 1835, one-quarter of Boston's 2,767 wells yielded water too putrid to drink. "The smell in many cases is extremely offensive, and I should think it probable that they have an injurious effect on the water of wells contiguous," an engineer's assistant noted that year. But odor and taste were relatively unimportant in an era when no one washed and everyone drank ale: it was disease, rather than malodorous wells, that prompted American cities to build reservoirs and aqueducts.

In this country's first century, American cities, like those of the Old World, were host to repeated epidemics, which killed more and more people as cities increased in size. No one knew how disease was transmitted, but there were two leading hypotheses. The imasmatics subscribed to the theory first put forth by Hippocrates more than two thousand years earlier: disease was caused by miasma, memorably characterised by one observer of the great 1665 plague of London as "a most subtle, peculiar, insinuating, venomous, deleterious exhalation arising from the maturation of the ferment of the feces of the earth." Contagionists believed that disease was passed from person to person, and that epidemics could be prevented by quarantining the city from foreign ships, halting trade. The miasmatics prescribed washing away the city's accumulated filth with plenty of clean water, a task of Augean proportions.

Considering what was clogging the streets, the miasmatics may well have been on the right track. Garbage was dumped in the alleys to be eaten by bands of feral pigs. Chamber pots (said to have been invented by the Sybarites, who were too lazy to go outside) were emptied out the door or into household cesspools. And when horsepower was produced by hay rather than fossil fuels, there was an enormous amount of manure to contend with. In 1780, when the population of Paris was about 600,000, city street sweepers removed 270,000 cubic meters of manure, cesspool collectors took a tenth of that, and the whole load was conveniently dumped into the Seine. The cesspool collectors would have accounted for only about a third of the feces produced annually by a city of 600,000, so the rest must have gone into the street. If Philadelphia was like Paris but one-tenth the size, the street sweepers collected about 30,000 tons of droppings annually, of which perhaps a fifth was human waste, while cesspool collectors managed 3,000 additional tons.

Yellow fever, which had entered the New World in 1647 with African slaves brought to Barbados, struck Philadelphia with a vengeance in 1793. Almost four thousand people died, and twenty-three thousand fled, effectively closing down the nation's capital and leading port for months. The disease skipped up and down the Eastern seaboard for the next few years, and in 1798 another thirty-eight hundred Philadelphians died and forty thousand people (about three-quarters of the population) left. The city's leading medical light, Dr. Benjamin Rush best known for his *Medical Inquiries and Observations upon the Diseases of the Mind* (1812), one of the first modern attempts to explain mental illness was a miasmatic. He proclaimed that yellow fever was caused by "the putrid exhalations from the gutters, streets, ponds and marshy grounds in the neighbourhood of the city," and that the cure was to cleanse the streets with fresh water piped from the Schuylkill River. Given the alternative closing the harbour most of the wealthy Philadelphians heartily endorsed his scheme.

Scholars were classically educated in those days they could read Frontinus in the original and were familiar with the Roman method of providing a city with cheap and plentiful water. But there were relatively few civil engineers in the United States at that time (and no engineering schools), so the city of Philadelphia contracted with an English civil engineer named Benjamin Latrobe to set up their water supply system. (In her interesting and artlessly named book *Bathrooms*, the interior designer Mary Gilliatt credits Latrobe with constructing the first bathroom in the United States to have a bathtub and sink as well as a toilet in a Philadelphia house, in 1810.)

By 1837, the city had spent about half a million dollars building a dozen miles of aqueducts and four reservoirs with a combined capacity of 22 million gallons, and laying over 98 miles of iron distribution pipes within the city. Nearly fifteen hundred houses had running water, and the streets were cleared of filth. "Blessed as we are above every other city of the Union, and perhaps not excelled by any in the world in the cheap and abundant supply of pure and wholesome water we now enjoy," burbled the *Annual Report of the Watering Committee* for the year 1836.

As yellow fever catalysed the construction of Philadelphia's first aqueducts and reservoirs, so cholera triggered New York's. Cholera entered the New World via Montreal in June 1832, popping up in town after town until it reached New York City a month later. By October, thirty-five hundred New Yorkers were dead and a hundred thousand had fled the city. Philadelphia, with its clean streets and plentiful supply of untainted water, lost only nine hundred lives. Work on New York's Croton Aqueduct system which included the Croton Dam, four reservoirs, and about 50 miles of tunnelled and bridged masonry aqueduct at a total cost of over $10 million began shortly thereafter.

The earliest city water supply systems were built by private companies, but as cities increased in size the proposed aqueducts began to rival those of the Romans, and the costs became so high that water supply systems were built by city governments. The funds were typically raised by selling bonds, and recouped by charging an annual fee of $10 or so for each householder's hook-up. There was no lack of takers: in 1849, nine months after the Cochituate Aqueduct opened in Boston, there were 10,851 private subscribers and 750 fire hydrants.

Every new subscriber to the water supply system increased wastewater flows, and the need for municipal sewer systems became increasingly apparent. Sewers were a necessity even before a municipal water supply system was built, when the water supply was from city wells: cellars would fill up during rainstorms and the streets became quagmires if the water had nowhere to flow.

In unsewered areas of a city, the streets were drained by digging depressions at critical intersections to collect the flow from the gutters. These drainage basins often became open cesspools, in which dead cats and dogs and garbage of all sorts piled up. In 1806, the British traveler John Melish wrote that New York City's drainage pools "accumulate such a collection of latent filth, that the steams of it are sometimes perceptible at two miles' distance."

In colonial times, the homeowners along a street would band together and dig their sewers collectively, laying a pipe directly to the nearest waterways. When kitchen sinks that could drain to a sewer replaced dry sinks, life was made much easier, so people went to great lengths to lay pipes. The expense was shared, and when latecomers connected to the sewer, they paid the builders for the privilege. The sewer network extended block by block, informally, until water supply systems increased the flow to the city and the jerry-built sewers began to back up. When Boston officially became a city, in 1823, one of the first acts of its government was to take over ownership and control of the area's sewers. The original lines had been laid for household drainage, but the city needed to build storm drains to clean the streets and empty the drainage basins. In the next fifteen years, the city laid scores of miles of sewers large enough to be entered for cleaning and artfully crafted of brick and stone. The cost of construction and maintenance was charged to connecting homeowners based on their assessed valuation, but since the householders couldn't be disconnected, the bills were rarely paid. Between 1823 and 1838, the city spent $121,109.52 on sewers, and less than a quarter of the cost was collected from the homeowners. Water supply and sewer systems allowed cities to clean the streets by washing the wastes that were clogging the streets into the nearest body of water. Sewers had not been laid with feces in mind, however; kitchen sinks were the household connection to the sewer, and Boston (for example) banned human wastes from its sewers until 1833. Popular acclaim for the flush toilet changed the central role of the sewers, the scent of the city, and the layout of residences as well.

There were no bathrooms before the advent of indoor plumbing: portable bathtubs were filled in the kitchen, washstands stood in each bedroom, and chamber pots hid under each bed. With a municipal water supply came bathrooms with fixed tubs that no longer had to be drained by hand, a plumbed bathroom sink rather than individual ewers, and toilets, which caught on fast once the sewer systems were built. Though the bowls didn't always clear and the flush was noisy, the toilet so outperformed the chamber pot that to use one was to desire one. The father of the flush toilet is thought to be Sir John Harington, a godson of Queen Elizabeth I. Sir John described his contraption in *The Metamorphosis of Ajax* (a pun on "a jakes," or outhouse), a book he published in 1596. His toilet had all the essentials but one: the pipe left the bowl in a straight line, so the water closet smelled like the cesspool below.

(Queen Elizabeth is said to have complained about the smell, at a time when people were relieving themselves almost anywhere.) That problem was remedied in 1775, when the British engineer Alexander Cummings patented a toilet with a siphon trap a pocket of water in a lazy-S pipe that kept the smell of the cesspool (or sewer) from wafting back into the bathroom. As soon as American cities had dependable water supplies and a sewer system, all the fashionable hotels installed bathtubs, sinks, and toilets, and the wealthier citizens followed suit. Rather than squatting on chamber pots and hauling buckets of water from pumps or carts, people were turning on the tap and enjoying the pleasures of a 5-gallon flush commode. In 1860, there were 6,500 flush toilets and 100 miles of city sewers in Boston, and the hotels used simply stunning amounts of water: Boston's Tremont House used over 25,000 gallons a day, and the Parker House over 20,000. By 1885, Boston had 100,000 toilets, 226 miles of sewers, and literally thousands of miles of pipes bringing wastewater to municipal sewer mains 8 feet in diameter.

Families with bathrooms use hundreds more gallons of water a day than farmilies without, and water consumption skyrocketed when bathrooms became commonplace. During the century in which Boston went from wells to flush toilets, daily percapita water consumption jumped from perhaps a dozen to nearly 100 gallons a day about the present rate. In the course of building water supply systems, a wholly new standard of living was born. Meanwhile, advances in microscope technology had revealed the presence of "animaliculae" in perfectly clean water, an alarming discovery. A pamphlet printed in 1845, shortly after the discovery of microorganisms in Boston's drinking water, showed an old Irishman peering into a microscope. "Well, did anybody ever see the likes of this!" he exclaimed to the crowd around him. "They call whiskey 'a droph o' the craiture.' But here's water that's a droph of a thousand craitures!"

Before these unsettling revelations, water had been judged on the basis of its scent, taste, and clarity, and there was a nearly universal belief that water that didn't offend the senses was above suspicion. The discovery that seemingly limpid water was actually teeming with tiny creatures was highly disturbing; the New York diarist George Templeton Strong noted on August 1, 1842, that people had "dreadful apprehensions of breeding bullfrogs inwardly." In spite of public interest and functional microscopes, the cause of disease wasn't fully formulated until 1880, when Louis Pasteur published *On the Extension of the Germ Theory to the Etiology of Certain Common Diseases* and replaced the concept of a nebulous malificent miasma with the theory that specific airor waterborne organisms caused specific diseases. In the next five years, medical researchers identified the microorganisms responsible for gonorrhea, malaria, typhoid, tuberculosis, diphtheria, cholera, and tetanus.

The discovery that would have perhaps the greatest impact on the nation's municipal water systems came in 1884, when the German pediatrician Theodor

Escherich put the feces of a cholera patient under a microscope in search of the organism that causes cholera. Those were the days when the world of science was an unwritten book, for the organism Escherich isolated, *Escherichia coli*, is hard to miss: they make up one-fifth to one-third of the weight of an average feces. Sure enough, further investigations showed that E. coli are also present in the feces of healthy individuals. Humans produce about two hundred billion coliform cells a day, making the amount of *E. coli* in waterways an ideal indicator of the water's fecal contamination. The odds of seeing a typhoid bacillus in a sample of water were slim, but a high count of E. coli was proof enough that the water was potentially infectious.

Sewering the city's wastes into the nearest body of water improved the health standards of upstream communities at the expense of those downstream. Typhoid fever, carried by feces, had become a significant cause of mortality: in 1880, the death rate from typhoid fever in New York was about thirty-two out of a hundred thousand; forty-two in Boston; fifty-eight in Philadelphia; and fifty-nine in Baltimore. Lawrence, Massachusetts, which drew its water from the polluted Merrimack River, had a typhoid mortality rate that remained eighty or more per hundred thousand from 1887 to 1894. When watertight household vessels were developed in pre historic times, water was clarified by storage, and also by filtration through porous receptacles or sand layers: in both cases, sediment was left behind. After many more centuries, it was found that filtering and sedimentation could be aided by the addition of precipitants and flocculants, which make the solids in the water clump together and sink: this discovery marked the beginning of water treatment. Sanskrit medical lore and Egyptian inscriptions from about 2000 B.C. are the earliest records of what is now known as primary water treatment. Alum, the flocculant most widely used to clarify water today, was first mentioned by Pliny in 77 A.D.; he described the use of both lime and alum for conditioning water. By the fifteenth century, the preparation of purified alum had reached manufacturing dimensions, and by 1767 the common people of England were purifying water by adding two or three grains of alum to a quart and allowing the impurities in the water to flocculate and settle out, after which the supernatant was filtered. In 1872, Poughkeepsie, which drank from the Hudson, built a sand filter and became the first American city with a filtered water supply. While the use of flocculation, sedimentation, and filtration of individual household and small industrial water supplies had been practiced from ancient times, the earliest use of flocculation for treatment of municipal water supplies occurred at Bolton, England, in 1881. From 1885 on, flocculation was widely used as a process preceding filtration.

Few American cities filtered their water supply in the 1890s; whereas twenty million Europeans drank filtered water, only about a hundred thousand

Americans did. Just a few years after *E. coli* was named and typhoid, cholera, and diarrhea were identified as waterborne diseases, filtration systems were shown to remove microorganisms from the drinking water. The result was a flurry of construction. Communities that built filterbeds had sharp declines in the rates of waterborne diseases, and the results were widely publicised. In 1893, the city of Lawrence built filters for its water supply, and the typhoid death rate fell from a high of 134 to between 20 and 30 people per 100,000. Innovations in mechanical filtration and sand filtration followed quickly, and dozens of articles and books plotted the death rate from waterborne disease against the type of city water supply.

In 1903, the waterworks pioneer Allen Hazen of Boston discovered that when a polluted city water supply was filtered, for every person saved from typhoid there were three saved from death by other causes, many of which had theretofore seemed unrelated to the water supply. This correlation between water supply and the mortality rate, known as the Hazen Theorem, brought filtration to the forefront of city agendas. No one had been very keen about drinking animaliculae in the first place, and the indication that they were associated with disease led most large American cities to build filtration systems for their water supply systems by 1910.

Chlorine was first used to purify water by William Cruikshank in England in 1800, though it wasn't applied to municipal systems until much later. Chlorination of public water supplies came to the United States in 1908, via the Chicago stockyards. The Union Stockyard Company had contracted with a private water supply company to provide the cattle with water from sewage-contaminated Bubbly Creek. The Bubbly Creek Filter Plant was built, and the water was filtered and cleaned by dosing it with copper sulfate, which kills algae; however, the cattle who drank the water didn't gain enough weight, so the stockyards started using city water again. The City of Chicago sued the Union Stockyard Company for poaching municipal water, and the Bubbly Creek Filter Plant then substituted chloride of lime (a source of chlorine) for copper sulfate; the cattle thrived, and the water was found to be cleaner than city water, in spite of its lowly origins. Later that year, Jersey City sued *its* water supply company for delivering tainted water; chlorine was added, and the courts ruled that the chlorinated water was legally and officially pure. Chlorination kills most microorganisms, and when chlorination is combined with filtration, the water supply becomes blameless.

The death rate from typhoid fever was the yardstick used by public health authorities to assess the public health of a community. As city water supplies began to be filtered and chlorinated, the average urban death rate from typhoid dropped from thirty-six per hundred thousand in 1900 to twenty per hundred thousand in 1910 to three per hundred thousand in 1935. Today, there are only a few dozen cases of typhoid fever a year in the United States. Maps of municipal water supply and sewer systems show the vascularisation of a city. Water is piped from pristine reservoirs or deep wells and diverted into hundreds of thousands of dwellings, where individuals turn on the tap and flush the toilets. The myriad of sewer lines join with many others to the sewer main, a great underground river with a daily flow that may be hundreds of millions of gallons, carrying the feces from millions of people to be discharged at a single point into the waterways. New York City's water supply system now includes more than 350 miles of aqueducts and 27 man-made lakes, which supply about 1.5 billion gallons of water a day approximately the same as the flow of the upper Hudson or of the Colorado at the Arizona-Colorado line.

When organic matter is dumped into a stream, it is broken down to its constituent parts and eaten by microorganisms, which require oxygen to metabolize their food. Since the aquatic ecosystem is fuelled by the organic matter in water, running water purifies itself, and the accepted wisdom among civil engineers and the public alike was that the waterways could be utilised for waste disposal almost without restriction. Water filtration and chlorination could protect the public from waterborne disease.

Industries that dumped their wastes into the water in the nineteenth and early twentieth centuries included beet-sugar processors, canneries, pulp and paper nulls, textile mills, dairies, meat packers, and tanneries, most of which laid a private pipe directly to the nearest body of water rather than connecting to the city sewers. Most of this industrial waste was simply organic matter (with the exception of effluvia from the tanneries and mills, both of which used an array of chemicals to break down fibres), and as long as there was enough oxygen in the waterways, a stream or river could assimilate them providing that the life in the water was not overwhelmed and asphyxiated.

But by 1912 there was enough concern about water pollution to warrant Congress's mandating a study of stream pollution. The Public Health Service funded a group of sanitary engineers, chemists, biologists, bacteriologists, and doctors to find out how much waste a stream could absorb, how much oxygen was required in order to break down different types of waste, and the efficacy of various treatment methods for industrial waste. One result was the "oxygen-sag curve," which defined how much waste a stream could assimilate. The dissolved oxygen of a stream could be measured, and the amount of oxygen needed to break down various wastes was characterised at specific times and temperatures. The dissolved oxygen was like cash in a bank, continually replenished by continuing deposits of aeration. The wastes were checks, which could be drawn against it.

The so-called Streeter-Phelps model, named for the two sanitation engineers that developed it, became the standard method of estimating the effects of industrial and human pollution on streams, and in some cases the oxygen studies were combined with bacterial surveys to form measures of stream sanitation and the adequacy of dilution. Industrial wastes were described in terms of their population equivalents that is, a unit of industrial waste was expressed in terms of the equivalent amount of wastes of a number of people. Unfortunately, oxygen limits are less relevant when inorganic or toxic components are added to industrial waste, a drawback that would not become apparent until the years following the Second World War.

Down the Drain

A revolution in chemistry took place in the years before 1940, and the urgency of war and the stimulus of postwar reconstruction put this new knowledge into practice. The technological revolution that followed the Second World War changed industry, agriculture, transportation, and communications. New industrial processes and products created literally thousands of new chemicals, and industrial effluent became increasingly toxic. Waste heat from power plants was also disposed of in the waterways, reducing the oxygen content of the water. Urban populations began to swell, and so did the flows of sewage. Mining increased in scale, contributing its own pollutants to the waterways, and agriculture adopted chemical fertilizers, insecticides, and herbicides, many of which washed into streams and rivers.The effects of all these new pollutants on the already simplified aquatic ecosystem was disastrous.

Better living through chemistry was the ethos of the postwar era, and the production of synthetic organic chemicals went from less than 1 billion pounds in 1940 to over 160 billion pounds by 1970; today there are some seventy thousand widely used synthetic chemicals, and about a thousand new chemicals are introduced each year. Many of these newly created substances were spread throughout the environment, and some of them became notorious.

DDT — dichloro-diphenyl-trichloroethane began to be marketed as an insecticide in 1944, the first of a family of chlorinated hydrocarbons that soon included 2,4-D (the herbicide Agent Orange), heptachlor, Dieldrin, Endrin, Aldrin, and Mirex, among others. A microscopic speck of these substances posed a substantial cancer risk; a pinch could kill you outright. When DDT first came on the market, its benefits were clear: it killed agricultural pests and thereby reduced the cost of food. Its drawbacks were less obvious.

Until DDT was released to the environment, natural processes were thought to dilute chemicals to the point where they virtually disappeared. It's a reasonable assumption, except that the biosphere tends to concentrate certain molecules. When DDT was swept into the waterways even in concentrations as small as 3 parts per trillion that is, roughly a drop for every 3 million gallons the life in water went to work. Zooplankton collected DDT at 40 parts per billion: the minnows that ate the zooplankton held 5 parts per million of DDT in their fat; and the fat of fish-eating birds, like eagles and pelicans, held concentrations of 25 parts per million. In a relatively short time, DDT could be found in organisms at 8 minion times the level in the water. At those concentrations, it interfered with calcium deposition on eggshells, and birds high on the food chain began laying eggs encased in membranes instead of shells. As a result, brown pelicans, ospreys, bald eagles, and other fisheating birds nearly disappeared.

PCBs — polychlorinated biphenyls were first manufactured in 1929, and were used as plasticizers and to produce paints and resins. Wonderfully inert, the new substance was widely used in electrical insulators. PCBs were known to be toxic, but since insulators were meant to be permanent installations, why worry? By the time it was understood that PCBs bioaccumulate and insulators are ephemeral from the earth's point of view, seals were having pups with fused fins. PCBs, as it turned out, are teratogens as well as mutagens and carcinogens, deforming the offspring of mammals at sublethal concentrations.

Some of the chemicals that backfired, like DDT and PCBs, were useful products, manufactured and introduced into the environment on purpose. Other exotic organic chemicals and heavy metals were dumped into the streams and rivers as unwanted by-products of industrial processes. Mercury, for example, was used as a catalyst in the production of caustic soda and acetaldehyde. In the 1940s, about 6 ounces of mercury were discharged into the waterways for each ton of caustic soda produced. Mercury was also used in the manufacture of the plastics urethane and vinyl chloride. The electrical industry used tons of mercury in batteries, silent switches, high-intensity street lamps, photocopying machines, and fluorescent lights. In 1969, the annual consumption of mercury in the United States peaked at over 3,000 tons.

It was thought that the mercury discharged into waterways settled to the bottom sediments and remained there, inert. But anaerobic bacteria in the bottom muck convert mercury into an organic compound called methyl mercury, which is readily concentrated by living organisms, including people. In Minimata, Japan, the local chemical plant had begun making acetaldehyde and discharging mercury into the ocean in 1932. Mercury discharges soared in the early 1950s, when the factory became the sole Japanese producer of a plasticizer that used acetaldehyde in its manufacture. The mercury was duly concentrated by the aquatic foodchain. In April 1956, there was an outbreak of a noncontagious illness of the central nervous system that would later become known as Minimata disease: the cats in town began to die, and the pediatrics ward of the Minimata factory hospital was crowded with brain-damaged children. Local fish were suspected as the source of the poison, and two years later the Kumamoto Prefecture banned the selling of Minimata fish.

Mercury was identified as the cause of the disease in 1959, and the factory reduced its mercury discharges but did not stop them altogether until 1968. In 1969, the factory owners were taken to court, and after four years of legal wrangling, they compensated more than three thousand victims of brain damage. Dioxin, a highly toxic carcinogen that affects the immune and reproductive systems of mammals at very low doses, is an industrial by-product with a twist: the molecule is created by the processes themselves, a side effect no one expected. Dioxins, along with other similar toxic molecules, are created whenever organochlorinated molecules are produced or destroyed. Dioxins occur in the chlorine bleaching process used to make paper, and in the incineration of such materials as PVC plastics or wood treated with pentachlorophenol preservatives. It is impossible to ban dioxins without banning the industrial processes that create them.

Some of the new chemicals released to the environment have been found to interact with the body's receptors for estrogen, which appear to be not so good at recognising what is a hormone and what is not a hormone. Nonoxylphenol polyethoxylates, used as surfactants in dishwashing liquids and toiletries, are estrogen mimics. Polycarbonate plastic, widely used in household containers and appliances, sheds bisphenol-A, an estrogen mimic so potent that 2 to 5 parts per billion are enough to induce increased hormone responses in reptiles, fish, shellfish, birds, and mammals. Endosulfan one of the most widely used pesticides acts like estrogen, and even the trace quantities left as residue on fruits and vegetables is enough to affect hormonal activity in many species, including human beings. DDT, PCBs, dioxins, atrazine, and some forty other compounds (many of them common household chemicals) have all been identified as hormone influences. Although these chemicals do not increase the levels of estrogen in an individual, the body perceives increased hormone levels. It is not clear how females are affected, but males of numerous animal species, from panthers to alligators, were found to have reduced sperm counts and smaller testes as a result of exposure to these chemicals, and some male fish develop genitals of both sexes. Worldwide, human sperm counts have dropped significantly since the 1940s, and a growing number of biologists believe that this may be due to environmental exposure to chemical estrogen mimics.

The chemical industry itself was responsible for most of the exotic new toxins in the waterways, but many other industries left their mark. Paper and pulp mills used massive amounts of sulfur, chlorine, sodium hydroxide, sodium carbonate, sodium sulfate, sodium chlorate, titanium dioxide, and aluminum sulfate, all of which were carried away in the wastewater. Textile production made heavy use of the waterways for waste disposal. When wool was carbonised, for example, strong acids were used to remove the cellulose impurities, followed by a soda-ash wash. Cotton fibres were coated with sizing to give them enough body to withstand the pressures exerted during weaving. Before 1960, the sizing was generally starch, and the desizing was done by enzymes, which had to be discarded into the waterways after each batch. Scouring of the fabric with hot alkaline detergents, caustic soda, and soda ash followed. And if the cotton was mercerised, sodium hydroxide and a sulfuric acid wash were added. Dying was a chemical nightmare, and the rivers running through mill towns were often rainbow banners of the daily changes in dye lots. Many of the synthetic fibres developed in the 1950s were water-repellent, so new sizings had to be developed. Instead of starch baths, polyvinyl alcohol, carboxymethyl cellulose, and polyacrylic acid baths were used to give the fibres body. Instead of enzyme desizing, sulfuric acid desizing was necessary for synthetic fibres like rayon and polyester, and the acid bath also ended up in the nearest stream, river, or lake. Molybdenum, chromium, vanadium, nickel, and other metals, many of them poisonous in trace amounts, were used in the manufacture of steel alloys. Pickling the steel, a tempering process, required significant amounts of cyanide. All these substances found their way into the waterways and some interacted synergistically, becoming more toxic together than they were individually.

By the early 1960s, the sluggish Calumet River (to give just one example), which empties into Lake Michigan, was receiving a daily dose of about 100,000 pounds of oil, 35,000 pounds of ammonia, 3,500 pounds of phenols, and 3,000 pounds of cyanide from the dense industrial complex around Chicago, Gary, and Hammond. The amount of organic oxygen-demanding waste added to the waterways increased as well. The food-processing industry used the waterways to dispose of the hair, offal, and blood from meat processing; the scum and whey from dairy products; the leavings from beet-sugar refining, brewing and distilling, canning and freezing. In the early 1960s, food processors in California's Central Valley dumped over a *billion* pounds of vegetable waste into the waterways each year during the canning and freezing season. Until the mid-1960s, the stockyards and packing plants of Omaha disposed of their wastes in the Missouri River, and a stretch of the river occasionally ran red with animal blood. Hair, entrails, and mats of congealed grease floated downstream and collected on islands and on the riverside. City populations swelled after 1940, and the sewer systems that had connected hundreds of thousands of toilets to a single main sewer now had millions of homes hooked in. More solids were discharged at a single point, and municipal wastewater treatment plants were built in many cities to prevent feces and toilet paper from fouling the beaches. Primary treatment plants removed much of the solids from the wastewater flow, but had little effect on the chemicals, including detergents.

Until 1946, people washed their clothes with soap, but as soon as synthetic detergents were sold they captured the market. Detergent molecules are long and asymmetrical: one end attracts dirt while the other end attracts water, which is why detergent lifts dirt from wet clothes. Detergent molecules come in two forms, straight and branched a difference that affects sudsing but not cleansing ability. The suds of the straight molecules of detergent break down quickly, while the suds of the branched molecules break down slowly if at all. In the early 1960s, detergent manufacturers waged an advertising battle over who had the longest-lasting suds, and the waterways lost. Suds began to appear in streams, rivers, and lakes, and in many areas great mounds washed up on shore. At the foot of Niagara Falls, the piles of discoloured detergent foam were 8 feet high.

Detergents also contained phosphate additives to soften the water and thereby improve the effectiveness of the detergent molecule. The environmental advocate Barry Commoner, in his 1971 book *The Closing Circle*, noted that between 1940 and 1970 the amount of phosphates in city wastewater increased from 20,000 to 150,000 tons per year. Phosphorus is a critical factor for plant growth in the water as well as on land. Wastewater, with its phosphorus-rich load of urine and feces, had long fertilised the receiving waters, and when the phosphates in detergents were added as well, the algae began to grow and grow and grow. Organic wastes provided so many nutrients so much food that the aquatic ecosystem became overwhelmed. On the bottom layers, the water was quickly depleted of oxygen, while anaerobic bacterial growth made the bottom muck stink and bubble. On the upper layers, the algae grew so splendidly on excess phosphorus that little else survived.

Continentwide, industries and cities were not the only polluters of the waterways. Agriculture adopted chemical fertilizers, pesticides, and herbicides; coal mines drained sulfuric acid into creeks and streams; and cyanide was used as a reagent to leach metals from low-grade ore.

As the demand for coal increased in the first decades of the twentieth century, the number and size of the coal mines grew. When coal surfaces are exposed, the sulfur in the coal comes in contact with water and air and forms sulfuric acid. As water drains from the mine, the acid moves into the waterways, and as long as rain falls on the mine tailings the sulfuric acid production continues, whether or not the mine is still operating. If the coal is strip-mined, the entire exposed seam leaches sulfuric acid. Strip mining was economical, but it rearranged the watershed, leaving the infertile subsoil on the surface and polluting streams. In 1939, West Virginia became the first state to require the reclamation of strip-mined land, and others followed.

By the late 1930s, it was estimated that American coal mines were producing about 2.5 million tons of sulfuric acid annually. Mine drainage was a particular problem in the Ohio River Basin, where twelve hundred operating coal mines drained an estimated annual 1.5 million tons of sulfuric acid into the waterways in the 1960s, and thousands of abandoned coal mines leached acid as well. The waterways were acidified wherever coal had been mined, though, and in Pennsylvania alone, mine drainage had blighted 2,000 stream miles by 1967. Farming, like mining, replaced labour with machinery and fossil fuels, and then added a chemical arsenal: fertilizer displaced land, herbicides displaced the cultivator, and insects were controlled by synthetic insecticides. Farmers had always added silt to the waterways, and in the postwar era they added chemicals as well.

Before the 1940s, farmers had used their wits to maintain soil fertility and contend with disease and pest cycles. (Prayers and rituals also played a part in good harvests. As recently as the 1900s, some Plains farmers sowed their wheat fields naked, under a full moon. The farmer's wife would stride through the furrows in front of her husband, who threw handfuls of seed on her rump by the light of the moon.) They rotated their fields between cool-season plants, like wheat, rye, flax, and oats, and warm-season plants, like buckwheat and millet; between dee-prooted and shallow-rooted plants; between broadleaf and grassy plants; between cash crops of grain and soil-building legumes, like hairy vetch.

Pesticides released farmers from their ancient battle against insects. Chemical fertilizers provided a cheaper and more concentrated source of nutrients than manure, and crop rotation was no longer necessary. Herbicides took care of the weeds that competed with crops for nutrients and light. With new highyield varieties, chemical helpers, irrigated fields, and oil-fuelled machinery, big harvests were nearly guaranteed. Corn, a heavy user of nitrogen, could be grown in a field year after year. For a short time, it seemed as though food production no longer depended on nature at all. Some of the drawbacks of industrialised agriculture became apparent quickly; others took decades to appear. The downside of pesticides showed up first. There was DDT's unnerving tendency to bioaccumulate, and then there was the high adaptability of insects, which lay lots of eggs. A few insects would invariably survive a toxic wipeout and produce resistant offspring, which would quickly repopulate the area. Pesticides also killed the beneficial insects that fed on the pests, removing natural checks from the ecosystem. And the poisons were washing into the waterways. In the 1960s, pesticides were reformulated to degrade more quickly, and pesticide use in the United States is currently between 250,000 and 375,000 tons a year about 2 or 3 pounds per person while the total crop losses to insects are approximately what they were before pesticides began to be used.

Chemical fertilizers provided no organic inputs to the soil, which began to lose tilth that is, the soil nutrients, texture, and ability to hold moisture declined. Without the addition of crop residues and animal manures, the organic matter in the soil was consumed and not replaced. As the soil compacted, chemical fertilizers had to be used in higher doses. Illinois farmers used 11,000 tons of nitrogen fertilizer per USDA unit of crop production in 1949; two decades later, they were using 57,000 tons to produce the same yield. The efficiency of nitrogen combining with the crop declined fivefold, and fertilizer nitrogen that did not enter the crops ended up in the ecosystem. In the West in particular, where water was impounded and diverted to grow food in former deserts, water used for irrigation would return to the rivers and streams laden with salts, minerals, and agricultural chemicals.

By the 1960s, the United States waterways had been thoroughly disrupted by dams, channels, diversions, and dredging. The watersheds were missing the wetlands and forests that once had cleaned the water; the riparian zone was often bare; and rangelands that had bled water into the groundwater now hemorrhaged soil into the streams. The aquatic ecoystems were stressed in every direction, the water cycle was unable to cope with the onslaught of chemicals and organic matter, and rivers and lakes began to die. Lake Erie, the shallowest of the Great Lakes which were called the Sweet Seas by the early French explorers was especially impacted by chemical and organic wastes. Along the north coast, silt and agricultural chemicals were swept into the water. Along the south coast, Buffalo, Cleveland, Toledo, and Detroit were dumping hundreds of millions of gallons of raw sewage a day, along with equally impressive amounts of industrial waste.

By the mid- 1960s, every major river flowing into the lake was grossly polluted. The Buffalo River was laced with oil, phenols, colour, organic matter, iron, acid, sewage, and various exotic organic compounds. Coliform counts reached 500,000 per 100 milliliters of water, and even sludge worms and leeches could not live in the industrially contaminated bottom sediments. The Detroit area added 20 million pounds of contaminants to Lake Erie every day, in a wastewater flow of 1.6 billion gallons. The Cuyahoga, which flows to the lake through Akron and Cleveland, ran chocolate brown on some days and rust red on others; it was so clogged with floating logs, oil, old tires, paints, and chemicals that the city of Cleveland declared it a fire hazard, and in 1969 it actually ignited, burning two railroad bridges in a toxic inferno of flames some 200 feet high.

The effects of Lake Erie's pollutants were soon spreading throughout the region. The insecticide Mirex, a chemical relative of DDT which was used to combat the spread of fire ants in the Southwestern United States, had been manufactured near Buffalo in the 1950s, and some had spilled into Lake Erie. It seemed secure in the bottom sediments until the early 1960s, when beluga whales in the St. Lawrence estuary started washing up dead of Mirex poisoning, mostly, though their organs were said to be so loaded with toxics that the belugas were technically hazardous waste. As it turned out, when the female eels of Lake Erie returned to the Sargasso Sea to breed, they swam there via Niagara and the St. Lawrence estuary. The eels' migratory reserves of fat were laced with Mirex and other chemicals. The beluga whales in the St. Lawrence estuary ate an average of 200 pounds of eels apiece per year, and many of the whales died. Mirex that was spilled into the waterways years ago is still cycling in the environment. By the 1960s, the watersheds surrounding Lake Erie had been debeavered, deforested, and plowed, the waters had been defished, the rivers to it had been dredged, straightened, and dammed. The flows of municipal and industrial wastes were unabated, and this Sweet Sea was sweet no longer. As it was in Lake Erie, so it was in waterways across the country. Nonetheless, as the fifties closed, the degradation of the natural environment was of little concern to the United States government. The business of America was business, and business was good. Late in his second term, President Dwight Eisenhower established a commission to set goals for America in the upcoming decade. The commission's report, delivered in November 1960, was 372 pages long, but water pollution and air pollution were covered in just five paragraphs. There were fifteen goals for national concern and action in the 1960s; the environment was not one of them.

And yet, with the growth of the suburbs, the public anxiety over nuclear testing, and the subsequent public education on chemical transport in the environment, the groundwork for the environmental movement had already been laid. In the postwar prosperity of the 1950s, blue-collar workers moved to the suburbs and bought cars. The automobile made wilderness activities like hiking, fishing, and nature photography more accessible, while pleasure boats and summer camps were no longer the province of the wealthy. As the population at large was rediscovering the delights of the natural world, the superpowers were testing their bombs. By 1951, the United States had detonated sixteen nuclear bombs and the Soviets thirteen; the following year, the British joined in with their first nuclear test. The explosions took place in remote locations, the results were shrouded in secrecy, and the newly formed Atomic Energy Commission would issue terse reports advising the citizenry that a test had occurred, and that the consequent radioactivity had been confined and was in any case harmless to the public.

One of the environmental movement's prophets was Rachel Carson, whose 1962 book Silent Spring described the effects of DDT on bird's eggs in lyrical prose, and expanded on the issue of chemical transport in the environment. If we continued to dump chemicals into the environment, Carson argued, we were likely to wake to a silent world. This book articulated the movement's central doctrine: the earth faced an ultimate "ecocatastrophe" unless immediate efforts were made to halt the deterioration of the environment.

The environmental movement was only beginning, but with a skilled and literate core the cause spread quickly. Pollution, which had formerly been seen as an inevitable by-product of an industrialised society, was redefined as a social problem: The more goods we produced, the more we stabilised the environment. In 1969, the same year the Cuyahoga caught fire, the National Environmental Policy Act became law, intending to "create and maintain conditions under which man and nature can exist in productive harmony." The first Earth Day, April 22, 1970, was a nationwide event, with the aim of educating the public about the environment. Organisations such as the Sierra Club, the National Resource Defence Council, Greenpeace, the Nature Conservancy, the National Wildlife Foundation, and the Audubon Society had huge gains in membership. The environmental movement began to capture the imagination of a nation, mounted successful political campaigns, and eventually changed the daily habits of millions of individuals.

In 1972, the Federal Water Pollution Control Act commonly called the Clean Water Act was passed, which ruled that water quality was to be improved without regard to cost and set as a "national goal" the ending of the discharge of pollutants into the waterways by 1985. Every city in the country was required to build and operate a wastewater treatment plant, and the newly formed Environmental Protection Agency (EPA) provided grants and technical assistance. The waterways were far from the only focus, though. Other environmental laws enacted included the Clean Air Act; the Resource Conservation and Recovery Act (aimed at hazardous waste); the Toxic Substances Control Act; the Federal Insecticide, Fungicide, and Rodenticide Act; the Safe Drinking Water Act; and the Comprehensive Environmental Response, Compensation, and Liability Act, which worked to clean up old hazardous waste sites. It was now the law of the land that environmental, aesthetic, and social impacts be considered along with economic impacts for almost any type of proposed land, water, or resource development. Rarely in history had the cry of the wilderness been heard so keenly, felt so deeply, and acted on so decisively by so many. The waterways, clearly, were saved.

The Clean Water Act set curbs on discharges of all types, but since most of the problem was perceived to have been caused by cities and industries, most of the money was focused on reducing pollutants from these sources. In the decade after passage of the Clean Water Act, the federal government disbursed billions of dollars in wastewater treatment-plant construction grants, and industries spent countless more billions on in-house solutions to clean up their effluent. Industry by industry and city by city, discharge reduction and source reduction were implemented, treatment plants were built, and toxic flows were reduced.

There were noticeable improvements. Fish kills declined, algal blooms began to disappear, and rivers and lakes around the country began to revive. The banning of DDT allowed the populations of ospreys, bald eagles, and brown pelicans to rebound, and recycling matured from a fringe activity to standard behaviour. One of the most obvious effects of the new environmental laws was that the coal-fired plants that once belched great gobs of black smoke were required to clean up their smokestack emissions. When the particulates were removed, the results were completely unexpected.

The larger particulates in smoke mostly ash are chemically basic. When the smokestacks were low and black smoke blanketed a city, the basic particulates neutralised the acids that form in the air from the nitrogen oxides and sulfur oxides released during combustion. The rain was dirty, but the pH was probably neutral; even London, known for its black shroud, had healthy trees and gardens. In the 1940s, taller smokestacks were built to raise the soot above the cities, and this is thought to have been the start of acid rain. The particulates still fell on the surrounding communities, but the acids circulated high and were swept northeast by the prevailing winds. When the Clean Air Act passed and the smokestacks of the Rust Belt were no longer allowed to blast wastes into the air, the largest particulates were removed by filters, while the sulfur oxides and nitrogen oxides went straight out the chimney. In the Northeast, the rain was soon as acidic as orange juice.

As long as the pH of soil is neutral or basic, the minerals and trace metals that lace the soil are bound in place. The soils of the Northeast were barely neutral to start with, and after three decades of increasingly acid rain, the soil was unable to buffer the onslaught. The metals that had been bound in the soil moved into solution, and fish began to die from metal poisoning. Amphibians, with their permeable skin, were directly affected by the acids, as were high-altitude trees, which were bathed in acid fog. By the early 1980s, trees on the summit of mountains in upper New York and Vermont had begun to die, and the mountain ponds in the region were often fringed with blanched corpses of salamanders and frogs. More equipment on every smokestack followed, but abatement technology was not the whole story. Energy conservation, improved combustion efficiency, changes in products and in the means of production soon contributed to cleaner skies and sweeter rain.

In 1987, fresh out of MIT with a master's degree in Technology and Policy in hand, I was persuaded that we could engineer our way to clean water as well. And so I went: to work on the last of the big-city wastewater projects: the Boston Harbour Clean-Up. Boston has one of the oldest sewer systems in the country, and one of the most sprawling. In the 1860s, the city built holding tanks on Moon Island in Boston Harbour, and released the sewage on the outgoing tides. At the time, this was a thoughtful step forward in sewage management, and Boston's innovative system was written up in civil engineering journals here and abroad. Over a century later, however, little had changed but the number of towns connected to the wastewater collection system. Two primary treatment plants, sited on small former islands south and north of the city, received the wastewater from forty-three communities, including over six thousand industrial dischargers and 2.3 million people. And the sludge was still being released on the outgoing tide every day. The brown slick was supposed to drift out to sea, but more often it ended up circling back to shore instead. And along with the sludge, the prevailing tides washed ashore thousands of pink plastic tampon applicators.

In 1982, the Quincy city solicitor was jogging along a beach near Boston and realised that he was running in turds. Within a year, the city of Quincy sued the Boston municipal commission that managed the wastewater treatment plants. Soon the EPA joined in the fray with another lawsuit, and it was ruled that a secondary wastewater treatment plant would be built and that sludge would no longer be dumped into Boston Harbour. The disposition of the sludge awaited the siting and construction of a sludge-pelletising plant, but the court had ordered that the scum be taken care of immediately. Before we get down to the details, though, we need to go deep into the innards of a wastewater treatment plant.

The tour starts in the lobby. Municipal funds for dressing up such unglamorous budget items are scarce, so wastewater treatment plants across the country are remarkably similar: the cheap linoleum and steel desks are as ubiquitous as the aquarium in the entryway and the whiff of digesting sludge from the works outside. The main that brings the raw sewage into a wastewater treatment plant is often enormous. In the United States, per-capita wastewater production is about 100 gallons a day, so city flows become epic. In Boston, it's greater than the combined flow of the three rivers the Charles, the Mystic, and the Neponset that feed into the harbour.

Wastewater is nearly all water, containing just 1 or 2 percent solids, including excrement and food, toilet paper and detergents, soaps, shampoos, cleansers, household hazardous waste, and oil and grease. The water is gray, and festooned with shards of toilet paper. A series of preliminary screens separate the socks, rags, sticks, and lumber from the sewage and protect the plant's pumps from debris. Next, the wastewater passes through chambers designed to slow the flow enough for grit to settle to the bottom. The grit is mostly sand and gravel, but also includes melon seeds, coffee grounds, cigarette butts (and, within a very short time, wiggly white worms); both the grit and the screened debris are usually turned into landfill.

The next stage of cleaning is done in giant rectangular openair primary sedimentation tanks, or clarifiers. A blade on rails moves slowly across the surface of each tank, skimming off scum and shunting it to thickening tanks. Seagulls whirl above, and dive down to fish out vegetables and tampon applicators, which litter the walkways between the tanks. Clarifiers hold the wastewater for an hour or two, which is long enough for the flow to separate into layers. The scum floating on top of the tanks includes grease, soap, skin, vegetable and mineral oils, some paper, wood, and cotton, along with most of the Band-Aids, condoms, and plastic tampon applicators people flush down the toilet. (These applicators are a wastewater engineer's nightmare. Aerodynamically designed, they slip through screens like a torpedo. In the 1980s, roughly fifty thousand applicators a day were arriving at the wastewater treatment plants in Boston. This translates to a flush rate for these items of about 40 percent, though Playtex, their principal manufacturer, claimed at the time that informal surveys indicated only 1 percent of their customers flushed the applicators.)

Meanwhile, the sludge mostly feces, toilet paper, and the heavier suspended solids sinks to the bottom of the clarifier tank, where it is scraped into a sump or a hopper and sucked out the bottom of the tank for further processing.

After primary treatment, the water has dropped about twothirds of the solids it carried, but it still contains some suspended, colloidal, and dissolved solids urine, for example. Before Boston built its secondary treatment plant, this water was chlorinated and released to the harbour. During secondary treatment, the dissolved solids in the water are eaten by a complex community of microorganisms that are already part of the flow. The temperature, oxygen level, and contact time are controlled to maximise the growth of a slimy, bulbous bacterial sludge, which looks like it would be ready to walk out of the refrigerator in different circumstances. The effluent itself is chlorinated and discharged to the waterways. Secondary treatment routinely removes about nine-tenths of the solids in wastewater, and 95 percent removal is not unheard of. Tertiary treatment, including filtration or chemical precipitation (in which chemicals are used to flocculate the dissolved solids), can remove over 99 percent of the solids in water. This effluent is drinkable (if you are so inclined), but tertiary treatment costs are generally considered much too high for such a marginal improvement, and only a handful of cities today have such systems in place. Ninety percent cleaner is pretty good, but secondary treatment is no cure-all. One-tenth of anything you put down the drain or the toilet or into the washing machine sneaks through the treatment plant and out to the waterways within hours.

Meanwhile, we've left the residuals still in progress back at the treatment plant. The scum is concentrated in the thickening tanks, where the water is drained off and piped to the head of the plant, leaving mostly grease behind. The sludge is thickened as well, and then these residuals are both piped to an enormous covered vat called a digester, in which the raw sludge brew a pungent, dark brown liquid the consistency of pancake batter -ages and ripens. It is kept warm, to encourage the growth of microorganisms, which consume digest the solids and give off methane. In most wastewater treatment plants

(Boston's included), this methane is collected and used to provide part of the plant's power. A digester usually holds the sludge for at least three weeks, so large treatment plants need several large digesters.

In a typical two-stage anaerobic digester, a half million gallons of sludge is kept well mixed during the first stage, maximising the contact of the organic matter with the microorganisms to encourage their growth. In the second stage, the mixer is turned off, and the sludge stratifies into four layers. Scum and odd chemicals rise to the top, and the comparatively clear supernatant below is piped back to the head of the plant for reprocessing. The actively digesting sludge burbles below all this, and the digested sludge with most of the organic matter consumed by bacteria sinks to the bottom, where it is pumped out of the digester to a holding tank. The grease in the scum is digested more slowly than the organic matter in the sludge, so a scum blanket breaks down steadily as it accumulates. The plastics in scum, though, are less tractable. Bacterial digestion usually reduces condoms to an innocuous-looking ring easily mistaken for a rubber band, but the plastic tampon applicators have great structural integrity, and in a few years a layer of them 6 to 8 feet thick builds up, and the digester has to be emptied and cleaned.

When bacteria and other microorganisms have consumed most of the organic matter in the water and the sludge is well digested, the result resembles composted cow manure mixed with water. The phosphorus in digested sludge once polluted the waterways, but it is a valuable fertilizer on land, while the organic material and the microorganisms build up the soil biota. Digested sewage sludge was routinely reused as fertilizer until the 1940s, when cheap chemical fertilizers became available. For the next forty years, sewage sludge was used as landfill, or else it was incinerated or dumped back into the waterways. But after the Clean Water Act curbed industrial contributions to sewers, sludge began to be used on land again. Seattle sprayed its sludge on forests to promote tree growth, while composted sludge from Maryland was used to fertilize the White House lawn. New York City continued to fill up barges and dump its sludge at sea, however, and Boston continued to release its sludge on the outgoing tides. Not until 1988 did Congress prohibit the release of sludge to the ocean, and the ocean dumping of sewage sludge in the United States finally ended in June 1992.

For two years from 1988 to 1990 Boston's scum was mine, but there was nothing to do with it. Landfill equipment slipped on the grease, so landfills would not accept it; incineration works poorly with a feedstock of grease and water; the fats were laced with sewage and useless for animal feed or soaps. Disposing of scum as a liquid was hopeless, so we made it into a solid. Every day, thousands of gallons of thickened scum were mixed with cement kiln dust, another waste product. The chemically fixed scum was a gray material encrusted with personal hygiene items. It behaved like a soil, mechanically, and eventually a few hundred thousand cubic yards of fixed scum was stored in three bulldosed hills. The rats and gulls would congregate there to eat; perhaps they liked the fats. My scum piles were ultimately incorporated into one of the berms designed to hide the new secondary treatment plant from view a fitting use, I thought, for the concrete evidence that people won't change their behaviour until they understand that their toilets are connected to the waterways.

Along with managing the scum, my job was to assess sludge quality to make sure that we met the EPA regulations for any sludge products that would be used on land. Nearly all the toxins and heavy metals in wastewater are transferred to sludge and scum in the wastewater treatment process. The chemical contaminants are not destroyed; some volatilise into the air, but most chemicals and heavy metals bind to either organic matter or grease, and end up in the digested sludge. Secondary treatment transfers 92 percent of PCBs in wastewater to the sludge, for example, along with 90 percent of the dimethyl phthalate and 70 percent of the cadmium. This means that digested sludge carries the fingerprint of all of the industrial discharges to the wastewater collection system.

Every city of fifty thousand people or more is therefore required by law to establish discharge limits for its industrial dischargers, a requirement that was initially ignored. When sludge started to be reused, though, municipalities got more involved in what arrived at the treatment plant. As soon as a city starts recycling its sludge, the industrial-discharge permit system tightens up. The inspectors pay closer attention, and an industry's furtive slips down the drain tend to decrease. A fortycity sludge survey in 1982 and the National Sewage Sludge Survey of 1988 both showed that cities that recycled their sludge had cleaner sludge, regardless of the number of industrial dischargers. According to the 1988 sludge survey, 13,458 wastewater treatment plants were generating nearly 6 million dry tons of sludge a year or roughly 50 dry pounds of sludge per person annually. Of that, over a third was land-applied, an amount that was steadily increasing.

This tale of the dwindling industrial contaminants in the wastewater grew even stranger, though, because not all the industrial contaminants in sludge get there because of industrial discharges. When rain falls on pavement, it sweeps oil, gasoline, and the detritus of tire treads down into the sewers. Households add detergents, soaps, cleansers, household hazardous waste, lawn-care products that would never be bought if people read the small print, and oil that weekend mechanics dump into the gutter. The water itself from the reservoir has its own baggage of metal and other contaminants.

With municipal-sludge quality reports from around the country piled on my desk, I began to realise that industries themselves are no longer directly dumping much down the sewers. Whether or not industrial discharge reports are reliable, sludge implacably reflects whatever contaminants are present in the waste stream; sludge doesn't lie. And most city sludge is remarkably clean. The waterways, however, are still polluted. What was dawning on me in the late 1980s because of local experience was beginning to show up nationwide. Water quality had markedly and rapidly improved in the 1970s, but by the mid-1980s a third of the nation's waterways were still assessed as unfit for fishing or swimming and so they have remained. About 40 percent of the lake acreage and 30 percent of the stream miles in this country are still polluted. Water quality has changed little in the last decade.

Over half the rain that falls on the continental United States roughly 35 billion gallons of water a day is made into wastewater by our cities and our industries, and our cities clean it reasonably well. The overwhelming majority of water pollutants are now contributed by big agriculture. In 1992, the Executive Office's Council on Environmental Quality attributed 6 percent of the impairment in streams and lakes to industrial pollution and a whopping 60 percent to silt and excess nutrients from fertilizer runoff. In 1992, the U.S. Public Interest Research Group, a nonprofit environmental organisation, estimated that toxic industrial contributions to the waterways and oceans totalled about 155,000 tons or less than 5 percent of what Detroit alone used to contribute every year in the 1960s. In the meantime, however, tens of millions of tons of topsoil wash into the streams, and farmers dump about 20 million tons of fertilizer on fields across the country, along with between 250,000 and 375,000 tons of pesticides. Agricultural subsidies which were designed in the 1930s, before the advent of chemical fertilizers and pesticides are at the heart of the problem. These subsidies, which are now being phased out, allow farmers to grow more crops than would be grown in a consumer-driven market, and in order to maximise their yields many farmers turned to the use of fertilizers and pesticides. Fortunately, organic farming techniques and biological (rather than chemical) pest management are gaining ground. Over a third of the nation's corn crop is now grown using low-till or no-till farming, in which the crop waste is recycled into the soil and earthworms create macropores, which reduce runoff and increase water uptake by the soil. The high costs of chemical fertilizers and pesticides and the premium that consumers are willing to pay for organically grown crops are changing the way many farmers manage their land.

But some types of pollution are far more difficult to manage. Groundwater pollution has proved to be surprisingly intractable; once a plume of contaminants is loosed into an aquifer, it is nearly impossible to remove. Likewise, airborne pollutants continue to rain down steadily from the skies even though the air has grown clearer. Remote ponds that appear to be pristine are polluted by assorted wind-deposited chemicals and metals, including mercury from incinerator smokestacks, DDT from Central American fields, and traces of other assorted hazardous wastes. Finally, we cannot undo what we have done. New hazardous waste sites are not being created, but the old ones are very costly to clean up; PCBs, once they have been released to the environment, cannot be taken back. Cleaning up industrial and municipal discharges has cost hundreds of billions of dollars, and a third of the nations's waterways are still polluted. Water pollution is clearly more complicated than we had realised, and discharge control has not solved the problem.

Consider this: after water leaves a reservoir, it takes hours or days to move through the pipes to a house or factory, where it's polluted; through a wastewater treatment plant, where it drops most of its pollutants; and out to the waterways or to the sea. After days running through the engineered system, it enters the natural water cycle for a decade or more, where it may run to the sea and eventually move to the clouds, be blown inland to rain onto forests and fields, run into wetlands and streams, or percolate down to the groundwater. And then, perhaps, it moves into a well or reservoir again. For each day that water flows in pipes, it might spend a decade or more in the natural world. And nature is the best cleanser at least, it once was.

This country's waterways have been transformed by omission. Without beavers, water makes its way too quickly to the sea; without prairie dogs, water runs over the surface instead of sinking into the aquifer; without bison, there are no groundwater-recharge ponds in the grasslands and the riparian zone is trampled; without alligators, the edge between the water and land is simplified. Without forests, the water runs unfiltered to the waterways, and there is less deadwood in the channel, reducing stream productivity. Without floodplains and meanders, the water moves more swiftly, and silt carried in the water is more likely to be swept to sea. The beaver, the prairie dog, the bison, and the alligator have been scarce for so long that we have forgotten how plentiful they once were. Beaver populations are controlled, because they flood fields and forests, while wetlands acreage decreases annually. Prairie dogs are poisoned, because they compete with cattle for grass, while the grasslands grow more barren year by year. Buffalo are generally seen as photogenic anachronisms, and alligators are too reptilian to be very welcome. But all of these animals once shaped the land in ways that improve water quality.

It is not only water quality that has been affected, though. Without these builders, the contour of the land has been smoothed, and many niches have disappeared; the base of the food chain has withered, and in the process the abundance and the productivity of the land has declined. In 1993, the National Biological Service was created by Secretary of the Interior Bruce Babbitt to gather biological data on public and private lands nationwide. Babbitt, a geologist by training and a former two-term governor of Arizona, was eager to bring ecosystem management to federally owned lands. In 1995, the National Biological Service completed the first ecological review of the United States. The study found that the extent and vitality of dozens of ecosystems throughout the country have suffered a sweeping, but largely unnoticed decline. In spite of gaps and uncertainties in the data, the researchers concluded that the information "portrays a striking picture of endangerment."

The tallgrass prairies, the oak savannas bordering the grasslands, and the old-growth, fire-managed deciduous forests of the Eastern United States are among the largest imperiled ecosystems, along with more than 100,000 square miles of longleaf pine forests that once covered the Southeastern coastal plain. The Eastern forests were cut, the Midwestern grasslands have nearly disappeared under crops, and the oak savannas have been degraded by fire suppression. The longleaf pine ecosystem has faced both types of decline: great swaths were cut down in the early part of the twentieth century, and without periodic fires the pines have been taken over by hardwoods. Cattle and sheep have destroyed much of the riparian edge in the grasslands, degrading the aquatic ecosystems. When a stream has a well-developed edge, it supports life that cleans the water. In the mixed-grass and short-grass prairies, cattle trample the edge of the stream. Without beavers to cut down or drown the older streamside trees, the cottonwoods become enormous, siphoning water from the flow. What used to be a live stream, with fish and a lush edge, becomes a barren gulch. The water table drops, erosion progresses, and in fifty or a hundred years the land becomes desert. In 1988, an Arizona Fish and Game Department report concluded that less than 3 percent of the state's original riparian zone remained intact, while, New Mexico has lost at least 90 percent of its riparian zone to grazing.

According to the National Biological Service's 1995 report, ecosystems that once covered at least half the area of the contiguous United States are now critically endangered. The face of the land has changed, and as ecosystems are fragmented and simplified, the species that rely on them for habitat are under increasing assault. The Nature Conservancy's 1996 Annual Report Card for United States Plant and Animal Species, a comprehensive assessment of the country's indigenous fauna and flora, has found that the mammals and birds are doing relatively well; however, flowering plants and freshwater species (like most mussels and many riverine fish) are not. Of 20,481 native species of plants and animals surveyed, about one-third are faring poorly. Over two hundred and fifty species are extinct or possibly extinct, over thirteen hundred are critically imperiled, eighteen hundred are imperiled, and over three thousand are considered vulnerable.

In the last century, twenty-one species of freshwater mussels and forty species of freshwater fish became extinct. Today, about two-thirds of the remaining freshwater mussel species in this country are at risk of extinction, along with about one-third of the species of amphibians and freshwater fish. All depend on rivers, streams, or lakes, which are becoming biologically poorer even as water quality improves. In their Annual Report Card, the Nature Conservancy attributes part of this to the long-term effects of dams and other water diversions.

When the Federal Energy Regulatory Commission (FERC) first issued fifty year operating licenses for dams in the 1940s and 1950s, rivers were managed to provide energy production, irrigated acreage, and city water supplies. Life in the river itself and along its banks and on its floodplains was not included in the equation. Dams cause periodic and drastic fluctuations in the channel flow, as water is released to generate electricity for peak power demands. Fish are cut off from their spawning grounds or killed in the electric turbines. Dams block the flow of nutrients, slow and heat the water, and use the kinetic energy that would reaerate the water to generate electricity. In the last twenty years, however, our view of a river has expanded to include recreation and aquatic habitat, and dams are beginning to be managed differently. In 1986, the FERC was required to give recreational and biological issues equal consideration with power generation whenever it renewed a dam's license or issued a new one. In many cases, so little of a dammed river's water had been dedicated to the aquatic habitat that a small decrease in power production resulted in huge gains for wildlife. On the Deerfield River in Massachusetts, for example, a 1996 agreement to decrease power generation by 10 percent is expected to result in a fifteenfold increase in the trout habitat.

The movement to give rivers back their form was spearheaded by whitewater enthusiasts, but soon reached many environmental and governmental organisations. In 1991, the Department of the Interior ordered that dams in Wyoming, Utah, and Colorado release water flows on the Colorado River in the spring and early summer to enhance river rafting, rather than using the water to generate power at peak usage times. On March 26, 1996, a week-long artificial flood was released from the Glen Canyon Dam, and the Grand Canyon was scoured by high spring flows after thirty-three years of low water. As a result, for the first time since it was dammed, the Colorado River is beginning to regain its natural flow. About eight hundred dams nationwide are up for relicensing between 1996 and 2010, and the waterways should eventually show the difference.

When more water is allocated to maintaining a river's ecosystem, there is, obviously, less water available for cities, industry, and agriculture. Demand, though, has proved to be more flexible than people had imagined. Agriculture consumes the largest fraction of Western water by far some 90 percent of it and installing more efficient irrigation systems and growing crops that require less water can reduce water consumption considerably. Industrial and municipal water consumption are both rate-sensitive, and when the price of water climbs the purchase of low-flow appliances increases, another way of reducing demand. The Army Corps of Engineers past masters of dredging and channelisation have also become more environmentally aware. In 1969, the Corps was branded Public Enemy No. 1 by Supreme Court Justice William O. Douglas for its environmentally devastating activities. In southern Florida in particular, the Corps projects were so destructive that they are finally now being undone.

The Kissimmee River originally meandered south for 140 miles before it drained into Lake Okeechobee. South of the lake lies the Everglades, a great sheet of water supporting a river of grass that once extended unbroken 150 miles to the southern tip of Florida. In 1928, a flood in south Florida drowned 2,750 people, and the Corps avenged their death by trammeling the Everglades. The sinuous curves of the Kissimmee were straightened into a 56 mile canal 175 feet wide and 30 feet deep. An immense earthen levee was built around Lake Okeechobee, and 1,400 miles of canals, levees, spillways, and pumping stations were built to free up over half of the total Everglades marshlands for agriculture. Land speculators, cattle ranchers, sugarcane growers, and other agriculturalists grew rich and powerful, but by the 1960s the river and lake were polluted. The southern end of the Everglades (1.4 million acres) was left wild, but too little water was allocated to maintain its ecosystem. By the 1970s, the former hordes of waterfowl had dwindled to flocks, agricultural runoff had overfertilised the plants and poisoned the animals, and the Everglades began to dry up. Today the Corps is working to free the Kissimmee from its dikes and locks, and to restore the wetlands that once cleaned the river. The original $372 million Corps of Engineers' project to restore 22 miles of canal into 43 miles of curvaceous Kissimmee has expanded into a $1.2 billion federal and state effort to restore the entire river. New water allocations to the Everglades are reviving the marsh, and people throughout the state are working together to ensure that Florida's paradise of grass is not lost. The Everglades is not an isolated case: in states across the country, the Corps now works in concert with the Audubon Society and other environmental organisations to manage water to maximise its benefit to wildlife.

After a hundred years of taking away the form of our waterways, we are starting to put some of the pieces back. This about-face will doubtless do much to help many aquatic species, but it may be too late to save most of the native freshwater mussels. The zebra mussel (*Dreissena polymorpha*) is an ancient resident of the Caspian Sea that spread throughout the European waterways in the last two centuries. In 1985 or 1986, biologists believe that a ship from Europe jettisoned a number of larval zebra mussels along with its ballast water into the waters of Lake St. Clair, which is tucked between Lake Erie and Lake Huron. The larvae of the zebra mussels, unlike that of the fishdependent American freshwater mussels, are free-floating. And while the

American mussels burrow in the bottom sediments, the zebra mussels, like their saltwater counterparts, cling with tough byssal threads to such surfaces as water intake pipes and ship hulls. The native freshwater mussels have been in decline for a century, and the zebra mussels are spreading to fill the ecological void. This alien has reproduced so successfully that in some areas the sparse populations of native mussels are encrusted with pistachio-size, elegantly striped zebra mussels. Each small mussel filters about a quart of water a day as it gorges on microorganisms, and the water in the Great Lakes and the Mississippi River is finally beginning to be cleaned by filtering mollusks again. The native mussels, though, are unlikely to survive the competition.

The success of the zebra mussel could bring the Endangered Species Act to its knees. This continent is rich in species of freshwater fish and mollusks endemic only to a relatively few small and scattered places. The Mississippi River drainage system is a major centre of endemism, as are the springs of the Southwest. As the populations of freshwater mussels Texas blind salamanders, tiger salamanders, Lahontan cutthroat trout, humpback chubs, Colorado River squawfish, and Devil's Hole pupfish dwindle, one can't help but wonder; Do we have to save all of them?

It's the wrong question. By focusing on the preservation of endangered species one after another, as if they were items in a catalogue, we are missing the larger, ecological picture. Without the ancestral complement of keystone species nature's engineers the path that water takes through the land, and the shape of the land itself, have been simplified. The central actors, which once numbered in the hundreds of millions and billions, are missing or scarce in most of their former range. Where do the salmon fingerlings hide and feed without the beaver ponds they coevolved with? How can the eagles prosper with no prairie dogs to eat? The tiger salamanders have no tunnels to hide in; the prairie chickens have no stage for their courtship dances. Without restoring the ancestral populations of engineers to at least some of the landscape, it seems unlikely that the supporting players will manage to survive. Back when I was managing scum, I learned that before people are willing to change the ways in which they impact the waterways, they have to understand how the system works. People will continue to flush plastics down the toilet, for example, until they understand that their toilets are connected to the waterways. People will continue to keep prairie dog towns and beaver colonies off their land until they understand that the pathways water takes through the land are changed by tunnels and wetlands and that their land will hold more water, and grow more grass, with these animals. Ranchers in the Southwest are still poisoning phreatophytes that steal water from the river, and the beavers that once made wetlands in that arid region and cut down the cottonwoods before they grew to be river-draining goliaths are forgotten. It seems unlikely that enough private landowners will allow either rodent species back on their land to make a difference in the nation's water quality. We have an enormous national commons, however. Over a fifth of the contiguous United States 626,000 square miles is publicly owned and federally managed by the Bureau of Land Management, the Forest Service, the Fish and Wildlife Service, and the National Park Service. In theory, both the prairie dog and the beaver could be restored on all that vast acreage with a stroke of the pen. But while it is well understood that degraded land pollutes the waterways, public land is managed with little thought to water quality. For the last century, timber production, grazing, and oil and mineral extraction have been the primary uses of our national commons. Beaver populations are controlled to maximise timber revenue, while net wetlands loss continues; riparian habitat is stripped by livestock, lowering water tables and degrading water quality; careless lumbering clogs the streams with silt.

The U.S. Forest Service manages about 265,000 square miles of forestland in the contiguous United States (and a lot more in Alaska), or about 8 percent of the forty-eight states. Although forests are water purifiers, that capability has been largely ignored. In the West, vast clear-cuts allow silt to clog mountain streams, affecting salmon fisheries and municipal water supplies. Forests, instead of cleaning our water, are fouling it. By logging in ways that preserve stream integrity. For example, by avoiding the logging of steep mountainsides and by limiting clearcutting public forestlands could begin improving stream quality rather than impairing it. The Bureau of Land Management, which manages our public grasslands, controls about 275,000 square miles of land (or about 9 percent) of the lower forty-eight. Some 23,000 livestock raisers lease this land and parts of Forest Service land as well, to graze a total of over 4 million cows and 2 million sheep. Some of the National Parks (which total about 75,000 square miles, or 2 percent, of the contiguous United States) support herds of buffalo, but the public grasslands of the Bureau of Land Management do not. There are 3,000 buffalo in Yellowstone National Park alone; in 1991, only an estimated 448 wild buffalo grazed on the bureau's 275,000 square miles. Aside from ripping up the water's edge, cattle and sheep are beset by predators in the forests and on the open range. Coyotes, bears, mountain lions, and wolves all prey on domesticated animals, and ravens pluck out the eyes of lambs. To keep the public lands safe for domesticated animals, the U.S. Department of Agriculture's Animal Damage Control Program destroys about 80,000 coyotes, 200 mountain lions, nearly 10,000 black bears, and 125,000 prairie dogs annually.

Without prairie dogs, beavers, or an intact riparian edge, the grasslands have barely survived. In 1991, the Executive Office's Council on Environmental Quality estimated that half the country's public rangelands were in poor or fair condition, while only 5 percent were judged to be in "excellent" condition-and because of recent efforts to reduce overgrazing and restore riparian habitat, this is the best condition the public rangelands have been for the past century. (Range in "fair" condition has between one-quarter and one-half the vegetation it should, while range in "excellent" condition has 75 percent or more.)

Grasslands need to be grazed, and cattle can be raised without degrading the land and water. Some Western ranchers are restoring riparian habitat by herding their cattle across the land to keep them from congregating streamside and by limiting foraging to specified parcels at particular times of the year. When sheepdogs protect a flock, predation drops sharply, and some ranchers are learning to live with more wildlife. Better still, buffalo are finally being raised for meat, and their populations are the highest they've been for a century. In 1993, a hundred and thirty thousand buffalo grazed, wallowed, and took good care of the water's edge; most of them lived on private ranchland, along with a hundred million cattle. In 1993, three private buffalo ranchers had larger herds than Yellowstone's. The country's largest buffalo herd is owned by media mogul Ted Turner, who recently also returned the prairie dog to his land. Buffalo are excluded from public grasslands because they are carriers of brucellosis, a disease that has little effect on buffalo but causes cows to abort their fetuses. As the buffalo population has increased, though, it appears that brucellosis is far less contagious than had been assumed, and that buffalo pose little threat to the cattle with which they share the range. Buffalo meat tastes a lot like lean beef. And since buffalo evolved on the prairies, they are far hardier than cattle and can be raised without antibiotics, hormones, or artificial growth stimulants. In the winter, they sweep the snow away from the grass with their massive heads and shoulders and eat snow for water; they survive temperatures that freeze cattle solid. Ranchers who raise buffalo sell their meat to speciality restaurants and some of the better meat markets countrywide; the price ranges from about $6 a pound for ground meat to $20 a pound for steaks, and so far the demand has outstripped the supply.

The popularity of the prairie dog, however, has not improved. Now probably no more than 1 percent of the area once covered by prairie dog towns is tunnelled. Although millions survive, their populations are vulnerable. The sylvatic plague is still endemic in prairie dogs, but the establishment of new colonies is restricted; without migration, Pasteurella pestis may get the best of the prairie dog. According to metapopulation theory, species are composed of networks of small, interacting populations that help to maintain one another: when populations decline, they can be rescued by healthy migrating neighbours. Healthy populations produce many migrants, thereby creating a positive-feedback loop resulting in a healthy population network. But if enough colonies falter, the network collapses. The sylvatic plague still depopulates prairie dog towns, and the migrants needed to repopulate them could become scarce.

Since the early 1980s, tensions have escalated between environmentalists and the spokesmen (mostly conservative politicians from Western states) for "endangered" loggers, mining companies, and ranchers. Large-scale logging, careless grazing, and corporate mining are still the main uses for our public land, but the environmentalists have succeeded in cordoning off nearly 52,000 square miles in eleven Western states almost 5 percent of the West as wilderness. The Northern Rockies Ecosystem Protection bill (not passed at this writing) seeks to set aside another 25,000 square miles in five areas, with wildlife corridors between them, which would put 1 percent of the country's old-growth forests under protection. Another 50,000 square miles of land about the size of Iowa has been taken out of agricultural production along the Mississippi flyway as part of the Conservation Reserve Program, which was set up in 1985. Ten years later, an estimated 83 million birds migrated south along the flyway, the biggest such migration in half a century.

There is the will to restore our land. We have just forgotten much of what is missing from it. The balance of nature that existed before we turned things upside down, and the richness and abundance of the land, were based on a few keystone species. What really matters are the numbers. In an area of 2.9 Million square miles, billions of prairie dog tunnels and countless millions of beaver dams and buffalo wallows are significant. With their removal went an ecological system that cleaned the water and enriched the land. In spite of our earnest engineering efforts, about a third of the waterways are still polluted, and the natural water cycle is still hugely simplified. But some of the filtering mollusks and the buffalo are coming back, and the prairie dogs and the beavers have both survived with their culture intact.

This land once had clouds of birds, dense herds of grazers, myriad shoals of anadromous fish and so it could again. It is time to restore the balance to our land and allow nature's engineers to do their work. If the prairie dogs and the beavers are allowed to re-establish their ancestral populations on public land, the dirt will fly, and the waterways will begin to regain their former pristine glory. On public land, at least, it is time for the beavers and the prairie dogs to come home.

Bibliography

Anderson, Mary P. and William W. Woessner: *Applied Groundwater Modeling: Simulation of Flow and Advective Transport.* Academic Press, San Diego, 1992.

Asaithambi, S. : *Economics of Ground Water Management in India*, Abhijeet, Delhi, 2008.

Bhakar, S.R. : *Ground Water Hydrology : Theory and Practice*, Agrotech, Delhi, 2009.

Bhave, P.R. and R. Gupta: *Analysis of Water Distribution Networks*, Narosa, Delhi, 2011.

Biswas Asit K. : *Integrated Water Resources Management in South and South-East Asia*, , Oxford University Press, Delhi, 2001.

Bourne, Peter G.: *Water and Sanitation: Economic and Sociological Perspectives* Academic Press. Orlando. 1984.

Brooks, D.: *Water: Local-Level Management*, Ottawa, International Development Research Centre, 2002.

Cullis, A. : *Rainwater Harvesting: The Collection of Rainfall and Runoff in Rural Areas,* London, U.K.: IT Publications, 1986.

Erik Nissen-Peterson : *Rainwater Catchment Systems*, UK: Intermediate Technology Publications, 1999.

Feder, Paul A.: *A Guidebook to Groundwater: Resources and Education Opportunities in the Great Lakes Region*, Great Lakes Commission, Ann Arbor, 1993.

Fetter, Charles Willard: *Applied Hydrogeology*, Merrill Publishing Company, Columbus, OH. 1988.

Geraghty, James J.: *Water Atlas of the United States*, Port Washington, NY, 1973.

Gischler, Christiaan E.: *Water Resources in the Arab Middle East and North Africa*, Cambridge, Eng, Middle East & North African Studies Press, 1979.

Gurjar, Ram Kumar : *Geography of Water Resources*, Rawat, Delhi, 2008.

Hemenway, Toby : *Gaia's Garden: A Guide to Home-Scale Permaculture*, Vermont: Chelsea Green Publishing Company, 2000.

Jat, M.L. and S.R. Bhakar: *Ground Water Hydrology : Theory and Practice*, Agrotech, Delhi, 2009.

Kanmony, J. Cyril : *Drinking Water Management : Problems and Prospects*, Mittal Pub, Delhi, 2010.

LeRoy, K. : *Sustaining Water Washington*, D.C.: Population Services International, New York, 1993.

McGarry, M.G.: *Matching Water Supply Technology to the Needs and Resources of Developing Countries,* United Nations. New York. 1987

Meenu Bhatnagar: *Groundwater Management : Sustainable Approaches*, Icfai Books, Delhi, 2012.

Morris, Mary E.: *Water Scarcity and Security concerns in the Middle East*, Abu Dhabi, Emirates Center for Strategic Studies and Research, 1998.

Narwani, G.S. : *Community Water Management*, Rawat, Delhi, 2005.

Nielsen, David M.: *Practical Handbook of Groundwater Monitoring*. Lewis Publishers, Chelsea, 1991.

Palanisami, K : *Groundwater Management and Policies*, Macmillan Publishers India, Delhi, 2008.

Premjit Sharma: *Agricultural Drainage and Water Quality*, Gene Tech Books, Delhi, 2007.

Singh, A.K.: *Environment and Water Resources Management*, Adhyayan, Delhi, 2006.

Thomas, Harold E.: *The Conservation of Groundwater: A Survey of the Present Groundwater Situation in the U.S.,* McGraw Hill, New York, NY. 1951.

Todd, David Keith: *Groundwater Hydrology*. John Wiley & Sons, New York, NY. 1959.

William, K. : *Irrigation Investment Technology, and Management Strategies for Development Studies in Water Policy Management 9*, Boulder, Colo Westview Press, 1986.

Index

P

R

S

T

V

W

❑❑❑